Anna Hadamus · Szczepan Piszczatowski ·
Małgorzata Syczewska ·
Michalina Błażkiewicz
Editors

Biomechanics in Medicine, Sport and Biology

Springer

Editors
Anna Hadamus
Department of Rehabilitation, Faculty of Medical Sciences
Medical University of Warsaw
Warsaw, Poland

Małgorzata Syczewska
Department of Rehabilitation
The Children's Memorial Health Institute
Warsaw, Poland

Szczepan Piszczatowski
Faculty of Mechanical Engineering, Department of Biocybernetics and Biomedical Engineering
Bialystok University of Technology
Białystok, Poland

Michalina Błażkiewicz
Department of Rehabilitiation
Józef Piłsudski University of Physical Education
Warsaw, Poland

ISSN 2367-3370 ISSN 2367-3389 (electronic)
Lecture Notes in Networks and Systems
ISBN 978-3-030-86296-1 ISBN 978-3-030-86297-8 (eBook)
https://doi.org/10.1007/978-3-030-86297-8

This Springer imprint is published by the registered company Springer Nature Switzerland AG
The registered company address is: Gewerbestrasse 11, 6330 Cham, Switzerland

Preface

This book contains 14 chapters which are full papers of selected presentations from the International Conference of Polish Society of Biomechanics (BIOMECHANICS 2020). This conference takes place in 2021, as the global pandemic outburst of SARS-COV-2 in December 2019 made the (planned) organization of the conference in September 2020 impossible. One year later, in September 2021, the traditional meeting in conference halls, face-to-face with invited lecturers, colleagues from various institutions and countries, representatives of commercial partners, is still impossible. Therefore, the meeting is in an online format. Of course, this is not a perfect solution, but the best we can offer. Long breaks in meetings and conferences are not beneficial for the scientific community, as the interchange of ideas, discussions, and critical comments push the research work forward. That was the reason behind the decision to organize the meeting this year — now in the new virtual world.

The restrictions related to SARS-COV-2 have influenced and continue to affect daily life but also the research. In many institutions, experimental research had to be suspended due to the closure of laboratories. It was not possible to invite subjects and patients to participate in research projects. In many universities, distance learning took teachers/researchers longer time than traditional learning. The researcher therefore had less time for scientific development. In some places, a big problem is the lack of funding due to pandemic time. Despite all these problems, the organizing committee of BIOMECHANICS 2020 received many interesting abstracts from various fields of biomechanics: sports biomechanics, clinical movement analysis, biomaterials, bioengineering, and many others. This book offers 14 full-length papers from different parts of modern biomechanics, which we hope can be of interest not only for the scientists but also for the students. We hope that the book will inspire you for further research.

We would like to thank all those who worked hard to create this book and the BIOMECHANICS 2020 conference: the organizers, reviewers, sponsors, and above all, the authors who contributed to the creation of this book.

On the behalf of the Board of the Polish Society of Biomechanics

Małgorzata Syczewska
Chairperson of the Polish Society
of Biomechanics

Lecture Notes in Networks and Systems

Volume 328

The series "Lecture Notes in Networks and Systems" publishes the latest developments in Networks and Systems—quickly, informally and with high quality. Original research reported in proceedings and post-proceedings represents the core of LNNS.

Volumes published in LNNS embrace all aspects and subfields of, as well as new challenges in, Networks and Systems.

The series contains proceedings and edited volumes in systems and networks, spanning the areas of Cyber-Physical Systems, Autonomous Systems, Sensor Networks, Control Systems, Energy Systems, Automotive Systems, Biological Systems, Vehicular Networking and Connected Vehicles, Aerospace Systems, Automation, Manufacturing, Smart Grids, Nonlinear Systems, Power Systems, Robotics, Social Systems, Economic Systems and other. Of particular value to both the contributors and the readership are the short publication timeframe and the world-wide distribution and exposure which enable both a wide and rapid dissemination of research output.

The series covers the theory, applications, and perspectives on the state of the art and future developments relevant to systems and networks, decision making, control, complex processes and related areas, as embedded in the fields of interdisciplinary and applied sciences, engineering, computer science, physics, economics, social, and life sciences, as well as the paradigms and methodologies behind them.

Indexed by SCOPUS, INSPEC, WTI Frankfurt eG, zbMATH, SCImago.

All books published in the series are submitted for consideration in Web of Science.

More information about this series at http://www.springer.com/series/15179

Acknowledgments

The International Conference of the Polish Society of Biomechanics "Biomechanics 2020" - a task funded under contract 756/P-DUN/2019 by the Polish Ministry of Education and Science allocated for science dissemination activities.

List of Contributors

Furkan Alemdar Department of Mechanical Engineering, Izmir Institute of Technology, Urla, Izmir, Turkey

Murat Bati Department of Mechanical Engineering, Izmir Institute of Technology, Urla, Izmir, Turkey

Liliya Batyuk Kharkiv National Medical University, Kharkov, Ukraine

Hamit Biçici Department of Mechanical Engineering, Izmir Institute of Technology, Urla, Izmir, Turkey

Onur Alp Buyukbayraktar Department of Mechanical Engineering, Izmir Institute of Technology, Urla, Izmir, Turkey

Romuald Będziński University of Zielona Góra, Zielona Góra, Poland

Emre Celik Department of Mechanical Engineering, Izmir Institute of Technology, Urla, Izmir, Turkey

K. N. Chethan Department of Aeronautical and Automobile Engineering, Manipal Institute of Technology, Manipal Academy of Higher Education, Manipal, Karnataka, India

Artur Cichański UTP University of Science and Technology in Bydgoszcz, Bydgoszcz, Poland

Adam Ciszkiewicz Faculty of Mechanical Engineering, Cracow University of Technology, Cracow, Poland

Muhammed Furkan Dasdemir Department of Mechanical Engineering, Izmir Institute of Technology, Urla, Izmir, Turkey

Hasan Göktaş Department of Mechanical Engineering, Izmir Institute of Technology, Urla, Izmir, Turkey

Jaroslav Hruby Institute of Forensic Engineering, Brno University of Technology, Brno, Czech Republic

Slawomir Grycuk Institute of Biomedical Engineering, Faculty of Mechanical Engineering, Bialystok University of Technology, Białystok, Poland

Paweł Ikoniak Faculty of Mechanical Engineering, Cracow University of Technology, Cracow, Poland

Mustafa Kara Department of Mechanical Engineering, Izmir Institute of Technology, Urla, Izmir, Turkey

Anatoly Khalin Kharkiv National University, Kharkov, Ukraine

Natalia Kizilova Warsaw University of Technology, Institute of Aeronautics and Applied Mechanics, Warsaw, Poland;
V.N. Karazin, Kharkov National University, Kharkiv, Ukraine

Tomasz Klekiel University of Zielona Góra, Zielona Góra, Poland

Karol Kluza Faculty of Mechanical Engineering, Cracow University of Technology, Cracow, Poland

Zdenek Krobot Department of Special and Combat Vehicles, University of Defense, Brno, Czech Republic

Agnieszka Mackiewicz University of Zielona Góra, Zielona Góra, Poland

Ryszard Maroński Institute of Aeronautics and Applied Mechanics, Warsaw University of Technology, Warsaw, Poland

Şenay Mihçin Department of Mechanical Engineering, Izmir Institute of Technology, Urla, Izmir, Turkey

Piotr Mrozek Institute of Biomedical Engineering, Faculty of Mechanical Engineering, Bialystok University of Technology, Białystok, Poland

Agnieszka Noszczyk-Nowak Wrocław University of Environmental and Life Sciences, Wrocław, Poland

Krzysztof Nowicki UTP University of Science and Technology in Bydgoszcz, Bydgoszcz, Poland

Monika Palmerska University of Zielona Góra, Zielona Góra, Poland

Kamila Pasik Department of Biomedical Engineering, Faculty of Mechanical Engineering, University of Zielona Góra, Zielona Góra, Poland

Jacek Rokicki (Deceased) Warsaw University of Technology, Institute of Aeronautics and Applied Mechanics, Warsaw, Poland

Marek Semela Institute of Forensic Engineering, Brno University of Technology, Brno, Czech Republic

Hadi Shirazi Department of Mechanical Engineering, Izmir Institute of Technology, Urla, Izmir, Turkey

Eda Subaşi Department of Mechanical Engineering, Izmir Institute of Technology, Urla, Izmir, Turkey

Metin Uzkut Department of Mechanical Engineering, Izmir Institute of Technology, Urla, Izmir, Turkey

Brad Parker Wham Center for Infrastructure, Energy and Space Testing, CU Boulder, Boulder, CO, USA

Bartłomiej Zagrodny Lodz University of Technology, Łódź, Poland

Biomechanical Aspects of in Vitro Fertilization

Liliya Batyuk[1], Anatoly Khalin[2], and Natalia Kizilova[3](✉)

[1] Kharkiv National Medical University, Nauki 2, Kharkov 61022, Ukraine
[2] Kharkiv National University, Svobody 4, Kharkov 61022, Ukraine
[3] Warsaw University of Technology, Nowowiejska 24, 00-665 Warsaw, Poland

Abstract. During the last decades, *in vitro* fertilization (IVF) became one of the most demanded reproductive technologies used for infertility treatment. Despite the significant efforts, the percentage of successful procedures remains moderate (<50%). It is shown, the percentage of successful IVF could be increased by a patient-specific embryo transfer based on the preliminary biomechanical and CFD analyses. A detailed review on different aspects of the IVF procedure is given. CFD simulations on the embryo transfer with tubular fluid and air bubble through a thin rigid tube (catheter) have been carried out. The following parameters were found to be the most influencing on the embryo transfer to the fundus: (i) the injection time IT, (ii) the distance of the catheter tip to fundus; (iii) the injected volume during the first stage; (iv) duration of the second stage; (v) the withdrawal speed at which the catheter is removed at the last stage; (vi) the volume replacement during catheter withdrawal. The IT, catheter load speed and cumulative shear stress over the particle during the IT were found the main prognostic factors of the IVF success.

Keywords: In vitro fertilization (IVF) · Embryo · Viscoelasticity · Damage · Computational fluid dynamics

1 Introduction

In vitro fertilization (IVF) is one of the widely known types of assisted reproductive technologies that includes artificial fertilization of the preliminary collected eggs, their culture during 4–6 days and further embryo transfer (ET) and its implantation into the uterus. The success of the IVF is estimated by the pregnancy rate (PR) that significantly depends on the age and a series of other factors of the women; it varies between PR = 39.6% at the age <35 to PR = 11.5% at the age >40 [1]. Mechanical factors play an essential role in cell divisions and embryo development [2], locomotion and invasion [3], differentiation [4], mechanotransduction [5], apoptosis [6], cell proliferation, gene and protein expression [7, 8]. Any variations in the mechanical environment influence the mechanical tension within the cytoskeleton of living cells that is a critical regulator of their biological function [5, 7], that will ultimately affect the respective functionalities. Mechanical inputs play a major role in regulating cell fate and function at the molecular level [9], and a cell's internal state is also reflected in its mechanical properties like elasticity, viscosity and dielectric coefficients [10–13].

A. Hadamus et al. (Eds.): BIOMECHANICS 2021, LNNS 328, pp. 1–16, 2022.
https://doi.org/10.1007/978-3-030-86297-8_1

During the last decade, the following aspects of IVF have been intensively discussed in literature [14–22]:

1) estimation of mechanical factors crucial for successful IVF and ET;
2) elasticity and viscosity of the embryo as indicators of its state and development;
3) microfluidic systems for embryo development at dynamical mechanical conditions;
4) development of robotic assisted equipment for IVF, embryo development and ET;
5) elaboration of the patient specific technique of ET.

2 Biofluid Mechanics of Reproduction

2.1 Mechanical Properties of Embryos

Each fertilized egg (oocyte) develops through the following stages: (I) morula, when cells have rapidly mitotically divided to produce a solid mass of 12–15 cells; (II) blastula or blastocyst that consists of an outer layer, an inner cell mass and a fluid-filled cavity; (III) blastocyst hatching (i.e. lost of the zona pellucid (ZP) which is a specialized extracellular matrix surrounding the developing oocyte within the ovary) followed by its free floating; (IV) implantation into the fundus. Structure and mechanical properties of the embryo at the stages I–IV differ significantly.

ZP (Fig. 1a) forms a multilaminar spherical shell of remarkably uniform thickness (5–10 μm) and a negligible flexural rigidity [23] that plays an essential role in both natural and *in vitro* fertilization [24]. ZP is a glycoprotein gel that can be modeled as viscoelastic Maxwell body [25]. The oocyte is free of residual membrane stress and after compression it recovers its initial shape with no residual deformations. The mechanical parameters of the oocyte and ZP can be measured in micropipette experiments (Fig. 1a). Embryo and oocyte stiffnesses are correlated, indicating that there may be a link between mechanics and viability [26–28].

After fertilization the mean Young's modulus of the ZP demonstrates a 1.8–2.4-fold increase preventing the oocyte from polyspermy [23, 29]. Starting from the 8-cell stage with cell divisions and gradual transition to the morula stage, ZP gradually softens again allowing further enlargement of the embryo [27, 28]. Therefore, the Young's modules of the ZP at different stages are known from the aspiration tests and some microfluidic experiments [3, 19, 22]. It was shown, the aged mice oocytes are significantly softer and demonstrate bigger relaxation times than the young oocytes [30]. Elastic modules (E) and apparent viscosities (η) of the embryo cells at the stages I, II, III were measured as $E = (0.9 \pm 0.4;0.4 \pm 0.2;0.4 \pm 0.1)$ kPa and $\eta = (166 \pm 81;77 \pm 37;40 \pm 15)$ $Pa \cdot s$ accordingly [31]. The embryos are very sensitive to mechanical properties of the substrate, i.e. the elasticity of the culture environment and the forces exerted over their surface. At the natural conditions, the stiffness of the uterine epithelium is ~1 kPa. Normally developed and damaged embryos are mechanically more strength than the damaged ones with blastomere fragmentation [32]. It was shown, the viability of the embryo at the 4-cell stage can be predicted with 90% sensitivity and 91% specificity based on the combination of mechanical (E_1, η_1) and two cell cycle parameters.

Based on the measurement data, several theoretical models of the embryo with ZP as an elastic incompressible half-space (half-space model), an elastic compressible bilayer

(2-layered model), and an elastic compressible shell (shell model) have been proposed [23]. The experimental studies with micropipette shown the layered and shell models are more precise. An equivalent rheological model as a modification of the Zener viscoelastic body (Fig. 1b) with, $E_1 = 0.3$ N/m, $E_2 = 0.12$ N/m, $\eta_2 = 0.59\ N \cdot s/m$ has been proposed in [33].

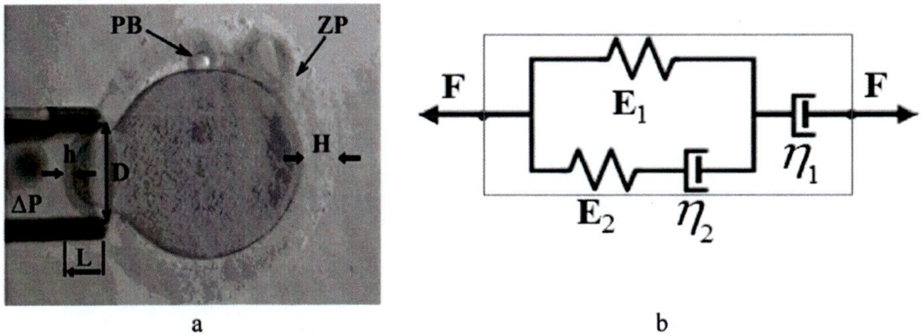

Fig. 1. Aspiration of the oocyte into a micropipetter (a) and its rheological model (b); H and h are the thickness of ZP outside and inside the pipette with inner diameter D, Δp is applied negative pressure; L is the aspiration length, $E_{1,2}$ and $\eta_{1,2}$ are elastic and viscous moduli.

2.2 Peristaltic Wave Propagation and Oscillatory Fluid Dynamics

Fluid dynamics aspects of the embryo development are connected with permanent fluid motion in the oviduct and uterus due to their smooth muscle contraction and beating of the ciliated epithelial cells [34]. Their importance was examined both experimentally and theoretically and different mechanical and electric properties of the tubular, ampulary and isthmic fluids have been shown [34, 35]. The embryo movement along the oviduct to the fundus is provided by ciliary beating of the epithelial cells; contraction of the smooth muscle; flows of the fluids secreted; thermotaxis due to the temperature differences (~0.02 C) in different parts of the uterus.

Cilia beating generate a continuous fluid stream that extends up to ~0.5 mm over the tips of the cilia. Therefore, it is highly probable that the embryos are not moved by a cilia stroke directly acting upon them but by the continuous fluid flow generated by the cilia beating with frequencies f = 5–20 Hz [36]. The embryo transportation can be conducted by both direct contact of the moving cilia with its surface and by the drag force ($\sim 1.1 \cdot 10^{-9}$ N) produced by the fluids accelerated by the cilia contractions [37]. The cilium beat has the following characteristics: (i) its rate is remarkably uniform, and (ii) the beat of a particular cilium and its adjacent cilium appears to be well coordinated and generate a wave with the period ~2.5 s and length $\lambda \sim 500$ μm [38]. The decreased ciliation and low cilia beat frequency are observed in women with obstructive tubal infertility [39]. The mechanics of viscous-dominated microscale flow *in vivo*, including time-reversal symmetry, drag anisotropy of slender bodies, and wall effects have been reviewed in [40]. Therefore, in the oviduct the embryo is subjected to the following dynamical forces [41]:

1) shear stress produced by tubal fluid flow;
2) compression by peristaltic tubal wall movement;
3) buoyancy due to differences in the density;
4) kinetic friction forces between the embryo and cilia.

All the factors lead to an average ovum velocity ~0.1 μm/s that can be increased to 3.8–6.8 μm/s by smooth muscle contraction of the walls. Fluid movements produced by wall peristalsis are similar to those induced by beating cilia and reach an average velocity of ~8.6 μm/s [36]. The motion of the embryo along the oviduct is not uniform. At some stages and conditions it could be accelerated or decelerated and even stopped by the active embryo-maternal interactions to provide the needed maturation of the embryo before delivering it to the fundus [42].

Based on the data that *in vivo* an embryo is naturally exposed to constant vibrations of ~6 Hz with the gradual increase to ~20Hz, a new approach to *in vitro* culture of human embryos with natural type mechanical micro-vibrations has been proposed [41]. It was shown, in the group of patients who underwent the IVT of the embryo(s) grown *in vitro* under mechanical vibrations the pregnancy rate was higher.

2.3 Mathematical Modeling of the Natural Embryo Transportation

In earlier analytical studies (1965–1985) simplified models has been elaborated, including infinite symmetric circular channels, inviscid fluids, small-amplitude oscillations, infinite wavelength, and elastic walls. Later studies accounted for the finite-length nonuniform channels, viscous fluids, non-steady low Reynolds number flows, finite wavelength and viscoelastic wall [34, 43, 44] and peristaltic flows of non-Newtonian viscoelastic fluids [45]. A 2D rectangle and tapered channel models with the walls oscillated sinusoidally with a phase shift $\Delta\varphi$, have been developed and solved based on the lubrication theory [46]. It was shown, the transport phenomena are strongly affected by $\Delta\varphi$. A two-layer flow with an inner layer through which the egg moves was coupled with an outer layer through which flow is assumed to be driven by the pressure gradient only [47]. The ciliary sublayer at the interface of the inner and outer layers was modeled by a distribution of volume forces produced by the cilia beating and muscle contraction.

A stochastic model of ovum transport based on Langevin's diffusion equation was proposed in [48]. The embryo motion was considered as one-dimensional random walks in a field of external force that included ciliary beating, smooth muscle contractions and the frictional force produced by the viscous fluid. The model predicted the leading role of cilia activity in fast transport of the embryo.

For Stokes flow, the grid-free method of regularized Stokeslets that restores the flow field from a distribution of regularized forces supported by the immersed boundary can be used. The models of cilia beating [34], peristaltic pumping of solid particles [34], and 3d swimming [49, 50] have also been studied based on the immersed boundary method (IBM). The coupling between the wall movement and the fluid flow can be described as a fluid-structure interaction (FSI) problem and solved by the IBM, finite element (FEM) and finite volume (FVM) methods. Nowadays the FSI modeling and FEM simulations is the most popular approach to the embryo transportation problem *in vivo* and *in vitro*.

2.4 Biomechanical Aspects of IVF

In vitro ET remains the most inefficient step in IVF procedure and limiting component to its success rate [51]. While fertilization in the laboratory is achieved at relatively high success rates (>90%), the maximum chance of pregnancy per cycle is <20% in healthy couples [52]. Only top clinics report PR >40%, while others have PR <10% and even in developed countries the number of clinics with a low PR is very high. Depending on the IVT technique, a pregnancy rate of 33.3% is considered as excellent transplantations and 10.5% as poor ones. Among other hypotheses, it was suggested that mechanical factors, such as catheter type, method of loading the catheter, placement of the catheter tip and injection speed (IS) may be the cause for the low PR. Therefore, methods to improve the effectiveness of ET are needed. The IS, the time between the end of the injection and the catheter withdrawal, the withdrawal speed are important fluid dynamics parameters that may have an effect on the result of the ET procedure [53].

Standard IVT procedure starts with a proper positioning of the patient, proper positioning of the catheter in the uterine cavity (Fig. 2a) under guidance of abdominal ultrasonography (USG). Then, the inner catheter with a soft tip preliminary loaded with the embryo(s) is placed inside the outer one by the laboratory technician or embryologist. The '3-drop technique' when a drop of medium (Embryo-Glue®) containing the embryo is separated from a preceding and following drops by air bubbles (Fig. 2b) is commonly accepted [54]. The air bubbles are visible by USG abd serve as markers for correct embryo's placement in the uterus. To minimize the risk of retained embryos, the catheter should be slowly withdrawn. Nevertheless, there is no unified point of view on the recommended times and speeds of the withdrawal.

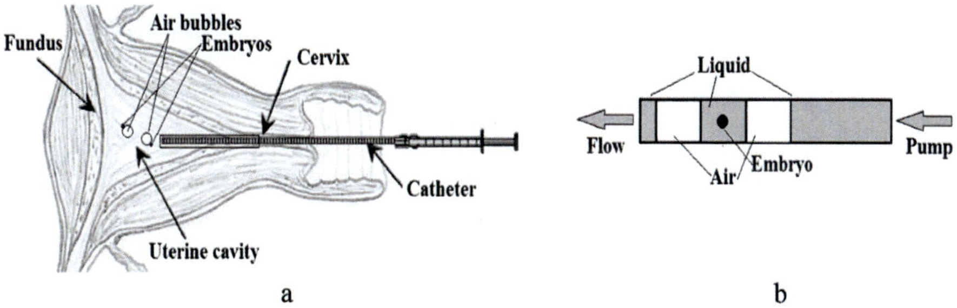

Fig. 2. Scheme of the embryo transfer (a) and the catheter tip loading (b)

2.5 Experimental studies on the uterine models

The IVT procedure with a loaded catheter (Fig. 2b) has been reproduced on a glycerin-filled rigid transparent model that imitated the inner cavity of the uterus (Fig. 3a). The model was installed on a plate inclined between 0–30° above the horizontal plane to repeat possible orientations of the uterus during the real ET procedure [55]. The transferred liquid was colored with a dye and its injection was recorded by a video camera. An example of the recording is given in Fig. 3b.

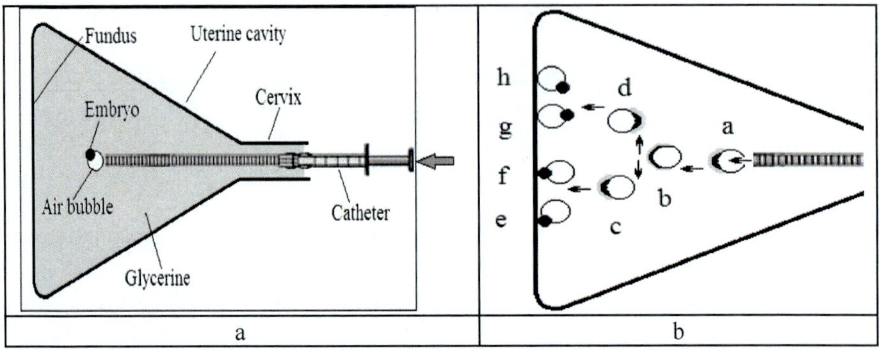

Fig. 3. A scheme of the experimental model for the laboratory simulations of ET (a) and possible trajectories of the bubbles with embryos (black) and transported liwuid (grey); e, f and g, h are successful and unsuccessful locations of the bubbles at the fundus.

As it was shown in the experimental studies on the model, its inclination angle influences the bubble formation and transportation of the embryo in the transferred fluid together with the bubble to the fundus via influencing the buoyancy forces [55]. Thus, the ET procedure must be performed at a patient-specific position in which the fundus location is exactly ahead of the buoyancy-driven trajectory of the bubble. In in vivo conditions this trajectory will also be influences by the active oscillatory forces produced by the soft tissues and fluid flows, that makes the decision making much more difficult. It was demonstrated that the air bubbles did not move after immediate ambulation after transfer in 94.1% of the cases [56] which means the fluid dynamics of the IVT is still poor understood.

It was confirmed that the underlying dynamics of the ET highly depends on the individual uterine anatomy (size, inclination angle), type and location of the catheter, the injection and catheter withdrawn parameters (timings and speeds); the relationship between all the factors are not well understood.

2.6 Catheter Tip Location

As it was reported in many papers, placement of the tip of the catheter in the middle of the uterine cavity resulted by higher PR [57]. In different studies the best location of the catheter tip is described as 0.5–2 cm from the fundus [57, 58]. When the catheter tip is placed close to the fundus (<5 mm), IR decreases. As it was shown in a detailed study based on 5055 US-guided ET [59], the PR is higher when the embryos are injected at some distance >10 mm from the fundus. Recently it was shown the embryo(s) must be released at 5–15 mm from the fundus [60]. The PR were 65.2%, 32.2% and 2.6% when the catheter tip was located at the distance <10 mm, 10–20 mm, and 20 mm, respectively [61].

2.7 Ejection Speed and Duration

Catheter ejection speed influences depth and placement of the embryo into the uterine cavity [62]. Nevertheless, the speed of injection is not defined in the public protocols;

it is just generally advised to do the ET very gently/slowly [58]. Despite numerous experimental studies on the best injection speed for successful ET, a large variation in PR rates 26–40% is observed among laboratory technicians [63]. The measurements of the speed and duration of the ejection conducted by 7 technicians revealed the differences from v = 2–4 m/s and T = 100 ms to v = 0.1–0.2 m/s and T = 750–1500 ms [63]. As it was shown in the experiments on the rigid transparent model of uterus [55], the injection speed significantly influences the air bubble formation and transportation together with the embryo(s). The minimal delivering time $t > 10$ s was proposed. An automated pump-regulated embryo transfer (PRET) unit has been proposed for standardization of the ET procedure [62, 63]. The pump applies the constant pressure to a syringe connected with the catheter and, therefore, provides a patient-specific constant speed of ejection. A randomized controlled trial conducted on 599 IVT procedures has shown, the PRET device generates a significantly smaller variance of the positioning of the embryo(s) into the uterine cavity that resulted in an ongoing PR = 21% in the PRET versus 17% in the manual ($p = 0.22$) transfer groups.

The patient specific hydraulic resistance of the uterine cavity must be accounted for but there are still no reliable experimental data on uterine resistance in ET [62].

2.8 Fluid Composition and Viscosity

The transfer media for ET are elaborated and manufactured by different pharmaceutical companies and have different compositions. One of the most important natural macro-molecule recommended to be introduced into the media is hyaluronic acid (HA) that is an implantation enhancing-molecule [64]. EmbryoGlue ® is a useful and available ET medium with a high concentration of HA.In the model studies the viscosity of the transfer media is often considered as equivalented to that of water, while the individual viscosity of human uterine fluid is not measured, though it is believed to be much higher. As it was shown on the laboratory model [55], better dispersion of the air bubbles with embryo(s) is achieved when the viscosity of the transfer media is close to that of the uterine fluid.

3 Mathematical Modeling of the Embryo Transfer

Individual anatomical features of the uterine cavity (size, geometry, inclination angle, presence of fibroids and other deformations) are important factors for the ET procedure [65]. Therefore, CFD simulations on the patient-specific models may provide the optimal biomechanical details of the ET for the individual patient. FSI modeling and CFD computations of ET have been carried out on a 2d model of uterus as a fluid-filled thin open ends channel with a rigid catheter inside (Fig. 4) [66]. The effects of uterine peristalsis on ET during and after injection have been investigated by the wall movements. The fluid flow was governed by incompressible Navier-Stokes equations. The peristaltic flow was generated by the transverse sinusoidal displacement of the walls with period T and phase shift φ in the form

$$H_1(t,x) = b\cos(2\pi(x/\lambda - t/T) - \varphi/2) - a,$$
$$H_2(t,x) = b\cos(2\pi(x/\lambda - t/T) + \varphi/2) + a.$$

The injection speed from the catheter tip was $U_{inj} = KU \sin(2\pi t/T_{inj})$, where U is the speed of the wave, K = const, T_{inj} = T/2 is the IT.

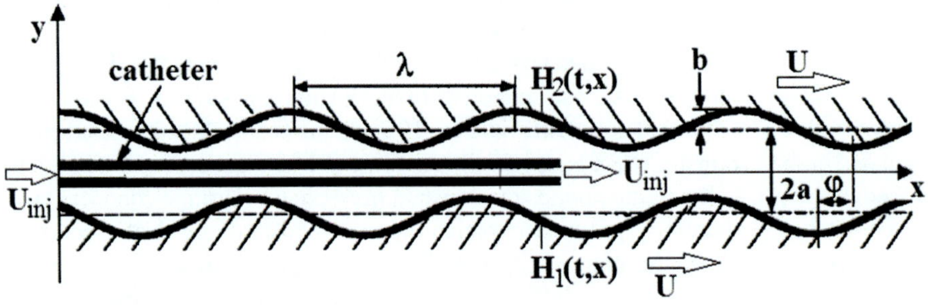

Fig. 4. The 2d model of ET into a channel with oscillating walls (from [66]).

It was shown, the displacement of the embryos as massless particles significantly depends on their location in the catheter (at y/R = 0, 0.25, 0.5 or 0.75, where R is the inner radius of the catheter). The central positions lead to the longer axial transfer of the particle, while any large distortion of the particle leads to the large amplitude radial displacements. The performed CFD studies revealed that higher injection speed of the particles located at y/R = 0 moves the embryo farther toward the fundus. The viscosity of the human uterine fluid was taken from the experimental measurements on animals $\mu = 1\ Pa \cdot s$.

Peristaltic flow in a 2d tapered tube model of uterus with oscillating walls and a rigid catheter along its axis was also studied [46]. The influence of the IT, the catheter distance from the fundus, the injected volume, the rest time, the withdrawal speed and the volume replacement generated by the catheter withdrawal has been modeled. Based on the CFD data, the risk function of the unsuccessful ET and the ectopic pregnancy as a linear combination of the average distance to the fallopian tubes, and location of the embryo relatively to the fundal wall was proposed.

A 3d uterine model as an axisymmtric tube with periodical varying cross-sectional area with a closed end (Fig. 5) was studied in [49, 50]. It was shown, the embryo trajectories are strongly affected by the damping out the peristaltic wave towards the end, and the trajectories of a particle of finite volume differ from the trajectories of massless particles. When initial location of the embryo was closer to the wall (far from the axis) the particles got scattered by the fundus and returned back to the catheter.

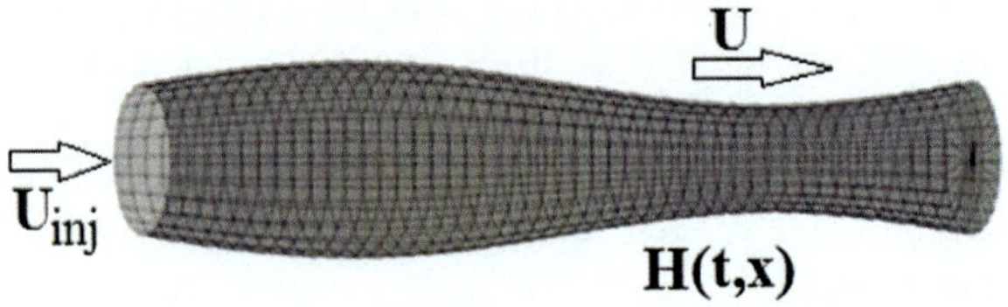

Fig. 5. 3d axisymmetric tube model of the uterine cavity (from [50]).

Based on the reviewed literature, the following essential biomechanical problems of IVF and ET can be formulated for improvement of the technique and increasing the PR:

1) Mathematical modeling of the ET along the inner tube between two air bubbles with embryo as a viscoelastic particle with different elastic and viscous coefficients (Fig. 2b);
2) Mathematical modeling of the ET between two air bubbles inside the model of uterus (both rigid and soft contracting walls);
3) Accounting for patient-specific geometry of the uterus (size, shape, anterverted/retroverted/midposition) that influences the buoyancy forces acting onto the air bubbles and the embryo movement between them;
4) Computer simulations of the ET with different size/location of the bubbles and medium aimed at proper embryo delivery and minimization the risk of retained embryos.

4 Mathematical Problem Formulation

Steady flow of two air bubbles (Fig. 6) of the length 7 mm in the polyethylene tube with D = 2R = 0.55 mm inner diameter with a column of the Embryo-Glue® fluid of the length 52 mm in between them is considered. The total length of the composition L = 83 mm. The IT is ~20 s, so the average injection velocity U_{inj} ~ 4 mm/s. For the comparative analysis the velocities U_{inj} = 1; 5; 20 mm/s have been chosen. These values correspond to the IT = 80; 17; 4 s. To avoid the influence of the entrance and exit flows on the flow parameters, the longer tubes of L = 15–30 cm will been considered.

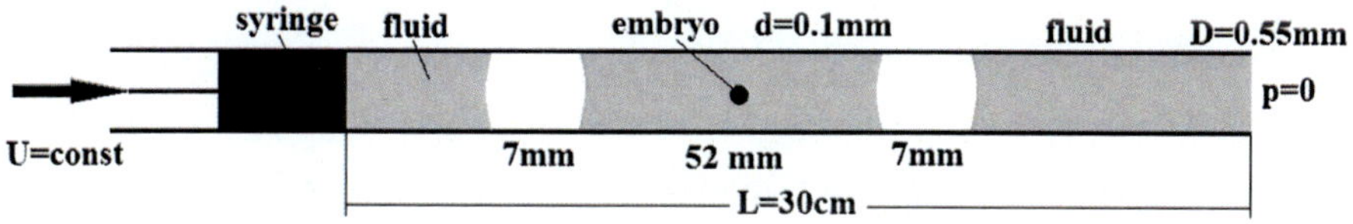

Fig. 6. A sketch of the modeled flow (the explanations are in the text).

The mathematical problem formulation for the multiphase flow is described by the steady Navier-Stokes equations [67]

$$\begin{aligned} &\nabla \cdot v = 0, \\ &\rho v \cdot \nabla v = -\nabla p + \mu \Delta v \end{aligned} \tag{1}$$

with the boundary conditions at the bubble surface S_b

$$n \cdot T = \frac{1}{Ca}\left(\frac{d\tau}{ds} + \frac{n}{r}\frac{dz}{ds}\right) - p_a n, \tag{2}$$

where n and τ are unit normal and tangent vectors at the point of the surface, p_a is the excess pressure inside the bubble, $T = -pI + (\nabla v) + (\nabla v)^T$,

$$z = 0: \quad T_{zz} = -p_{in}, r = 0: \quad T_{zr} = -0, r = 0 \cup S_b: \quad n \cdot v = 0. \tag{3}$$

Solution of the problem (1)–(3) has been obtained by the finite element method and Ansys Fluent 19.2 for the multiphysics flows at low Reynolds numbers $Re = 10^{-3} - 10^{-1}$. The refined mesh with inflations to the walls, interfaces and injected particles have been generated. The pressure, velocity and shear stress fields in the tube have been computed. The sample spherical elastic particles with d = 0.05 mm (smaller embryos) and d = 0.1 mm (larger embryos) have been located in the middle section of the liquid column (Fig. 6) at the radial coordinates r/R = 0; 0.25; 0.5; 0.75 (Fig. 7) and their trajectories have been computed.

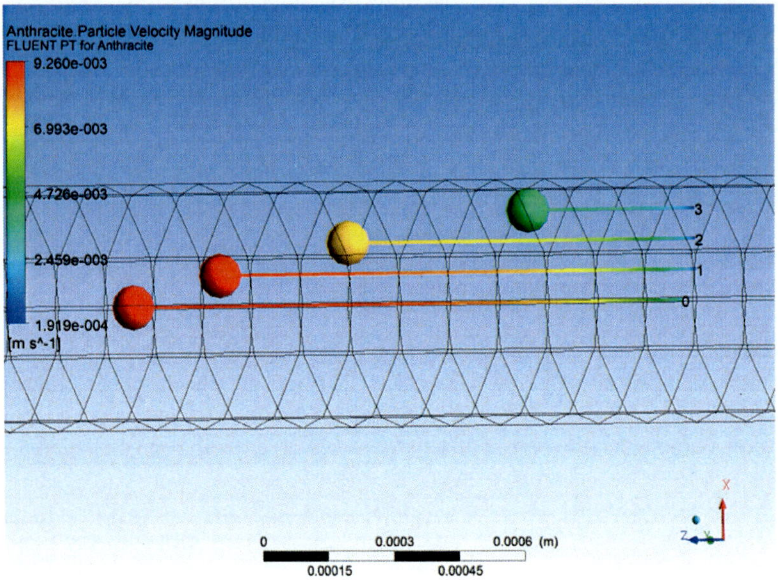

Fig. 7. Velocities of four sample particles at the time t = 2 s.

5 Results

One of the most useful approaches for the cell damage estimation is computation of the history of the damaging factors (pressures, pressure oscillations, shear stresses) during the cell movement [68–70]. It means, the listed factors must be computed along the trajectory of the sample particles in the tube. Some cells that move faster can experience higher mechanical forces onto their surfaces during a shorter travel time, while the cells that move at lower injection rates will experience the lower mechanical stresses during the longer time. In that way, the optimal injection speed with the lowest damage factor could be determined from the computer simulation data.

This approach has been validated for quantitative estimation of the red blood cell (RBC) damage in the apparatus of the extracorporeal circulation, through the stenosed mitral valve and valve prostheses. Various mathematical models have been used to correlate the mechanical forces and exposure time to the degree of the RBC damage. The most popular models are (i) average shear stress model [70], (ii) temporal differential model [71], and (iii) total differential model [72].

Based on this approach, the non-dimensional damage accumulation factor (DAF) can be computed by a simple formula

$$\mathrm{DAF} = \frac{1}{\mathrm{IT}\cdot\Sigma\tau*}\int_0^{\mathrm{IT}}\int_{\Sigma}\sqrt{\tau_{12}^2+\tau_{13}^2+\tau_{23}^2}\,\mathrm{d}\Sigma\mathrm{d}t, \tag{4}$$

where Σ is the surface of the particle, $\tau*$ is the critical dimensionless shear stress factor (at $\tau > \tau*$ the surface of the particle will be damaged), τ_{ik} are the components of the shear stress tensor computed at the surface of the particle.

For the more detailed quantitative estimations of the damage, the integrals of normal forces (pressures), pressure oscillations, and deformations of surface may be included into DAF with corresponding weighting coefficients. As the threshold value the critical wall shear stress $\tau* \sim 120$ mPa for mouse embryos [73, 74] is used here.

DAF computations have been carried out by gradual iterations with different time steps Δt for different injection speeds and data collection on τ_{ik} at each time step. When the particles were located closer to the wall of the tube, the shear stresses at their sides were non-symmetrical due to the parabolic-type flow profile of the fluid. This asymmetry generates rotation of the particle in the direction to the wall that produces the Magnus forces governing the particle towards the axis of the tube. In the concentrated suspensions such motion is suppressed by the repulsive forces, while in the case of single embryos there are no returning forces and the motion of the particles is more complex.

The computed DAF values are presented in Fig. 8a,b. The solid straight line corresponds to the damage threshold $\tau*$. As it follows from the results, the very slow injection produces low shear stresses which might be slightly over the critical value due to longer travel time, but for the computed values T = 83 s the increase in very small (DAF = 1.04). The faster injection produces higher DAF indexes DAF ~1.4 that might be very influencing for the embryos. Based on the interpolations of the DAF values computed for three injection velocities, the safe regions of injection can be found for different embryo diameters. Since, location of the particle along the radius of the tube is unpredictable, the worst case with r/R = 0.5 must be chosen. It gives us optimal injection velocities $U_{\mathrm{ing}} \sim$ 3–6 mm/s for the smaller particles, that gives DAF = 0.85 and IT = 18–27 s. For the bigger embryos the injection velocities v ~ 4–6 mm/s that gives DAF ~0.95 and IT = 18–21 s can be recommended.

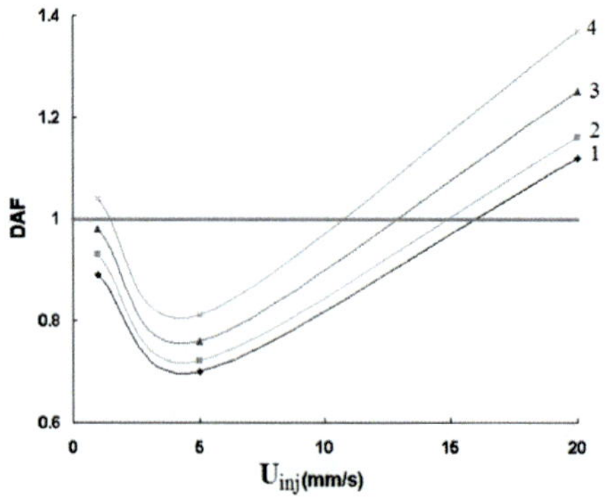

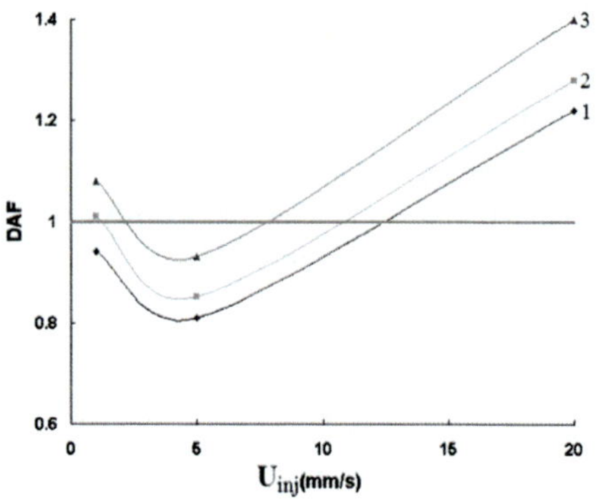

Fig. 8. Dependences DAF(U_{inj}) computed for the particles with d = 0.2 mm (a) and d = 0.1 mm (b); the curves 1–4 correspond to r/R = 0; 0.25; 0.5; 0.75.

6 Conclusions

The literature review revealed the following biomechanical aspects of the IVT procedure:

1) Viscoelastic properties of the 4–6 day old embryos [3, 19, 22, 31, 75];
2) Existence of the peristaltic waves and oscillatory fluid dynamics [34–42];
3) Catheter shape and tip location [55–61];
4) Ejection speed and duration of the procedure [62, 63];
5) Fluid composition and viscosity [55, 64];
6) Importance of the individualized procedure based on preliminary mathematical modeling and artificial intelligence approaches [76].

The steady viscous flow of three columns of the Embryo-Glue fluid with the embryo inside the middle column separated from a preceding and following columns by air bubbles of given volumes through a circular tube is studied. This problem has never been studied before in application to the ET procedure. The material parameters of the fluids, elasticity and damage shear stress values have been taken from published experimental data on zebra fish, mouse, and human embryos. A series of computational sets for three injection velocities, two embryo sizes and four initial locations of the particles have been obtained. The velocity fields, pressure distributions, pathlines and shear stresses have been computed. The mean and maximal values of the hydrodynamical parameters strongly depend on the injection velocity, particle size and location according to the tube's wall. The obtained results are in qualitative correspondence with experimental and CFD results [49, 50, 65, 66] obtained on the massless particles and small rigid particles in the tubes with rigid straight and waved surfaces.

A novel DAF paratemer for estimation of the optimal patient-specific injection speed based on known size, elastic parameters and developmental stage of the embryo is proposed. The DAF is computed as the shear stress experienced by the embryo's surface embryo during its injection. The approach is well known and validated for the RBC damage in the extracorporeal systems, valve prostheses or stenosed/calcified valves. For the ET procedure the approach has not been studied yet. It was shown, the numerical computations for three different injection velocities gives enough data for the interpolation of the dependencies DAF(U_{inj}) and estimation the limits for optimal U_{inj} for known embryo size and properties. Fast injection may produce high DAF due to high shear stresses especially when the embryo is closer to the wall, while the slow injection may also give higher DAF values due to longer IT of the embryo even at the low shear stress conditions.

The obtained results are based on an oversimplified elastic model of the embryo and the values measured on the mouse blastocysts only. More reliable data can be obtained on the realistic viscoelastic models of the embryo. The computed results also need validation by the experiments on human embryos. Other potential damaging factors (pressures, pressure gradients, variations in time, deformation of surfaces, etc.) can also be accounted for in (4) that will be done in future studies.

References

1. WebMD. https://www.webmd.com/infertility-and-reproduction/guide/in-vitro-fertilization
2. Krupinski P., Chickarmane, V., Peterson, C.: Simulating the mammalian blastocyst - molecular and mechanical interactions pattern the embryo. Comput. Biol. **7**(5), e1001128 (2011)
3. Yanez, L.Z., Camarillo, D.B.: Microfluidic analysis of oocyte and embryo biomechanical properties to improve outcomes in assisted reproductive technologies. Mol. Hum. Reprod. **23**(4), 235–247 (2017)
4. Huang, S., Ingber, D.E.: The structural and mechanical complexity of cell-growth control. Nat. Cell Biol. **1**, E131–E138 (1999)
5. Chicurel, M.E., Chen, C.S., Ingber, D.E.: Cellular control lies in the balance of forces. Curr. Opin. Cell Biol. **10**, 232–239 (1998)
6. Janmey, P.A.: The cytoskeleton and cell signaling: component localization and mechanical coupling. Physiol. Rev. **78**, 763–781 (1998)
7. Wang, J.H.-C., Thampatty, B.P.: An introductory review of cell mechanobiology. Biomech. Model Mechanobiol. **5**, 1–16 (2006)
8. Huang, A.H., Farrell, M.J., Kim, M., Mauck, R.L.: Long-term dynamic loading improves the mechanical properties of chondrogenic mesenchymal stem cell-laden hydrogels. Eur. Cells Mater. **19**, 72–85 (2010)
9. Engler, A.J., Sen, S., Sweeney, H.L., Discher, D.E.: Matrix elasticity directs stem cell lineage specification. Cell **126**, 677–689 (2006)
10. Suresh, S., et al.: Connections between single-cell biomechanics and human disease states: gastrointestinal cancer and malaria. Acta Biomater. **1**, 15–30 (2005)
11. Xu, W., et al.: Cell stiffness is a biomarker of the metastatic potential of ovarian cancer cells. PLoS ONE **7**, e46609 (2012)
12. Batyuk, L., Kizilova, N.: Modeling of dielectric permittivity of the RBC membrane as a three-layer model. In: Development trends in Medical Science and Practice, pp. 18–37. Baltija Publishing, Riga (2018). https://doi.org/10.30525/978-9934-571-31-2
13. Batyuk, L., Kizilova, N.: Dielectric properties of red blood cells for cancer diagnostics and treatment. AS Cancer Biol. **2**(10), 55–60 (2018)
14. Foo, J.Y., Lim, C.S.: Biofluid mechanics of the human reproductive process: modelling of the complex interaction and pathway to the oocytes. Zygote **16**, 343–354 (2008)

15. Muglia, U., Motta, P.M.: A new morpho-functional classification of the Fallopian tube based on its three-dimensional myoarchitecture. Histol. Histopathol. **16**, 227–237 (2001)
16. Kim, M.S., Bae, C.Y., Wee, G., Han, Y.M., Park, J.K.: A microfluidic in vitro cultivation system for mechanical stimulation of bovine embryos. Electrophoresis **30**, 3276–3282 (2009)
17. Beebe, D., Wheeler, M., Zeringue, H., Walters, E., Raty, S.: Microfluidic technology for assisted reproduction. Theriogenology **57**, 125–135 (2002)
18. Swain, J.E., Smith, G.D.: Advances in embryo culture platforms: novel approaches to improve preimplantation embryo development through modifications of the microenvironment. Hum. Reprod. Update. **17**, 541–557 (2011)
19. Wheeler, M.B., Walters, E.M., Beebe, D.J.: Toward culture of single gametes: the development of microfluidic platforms for assisted reproduction. Theriogenology **68**, S178–S189 (2007)
20. Krisher, R.L., Wheeler, M.B.: Towards the use of microfluidics for individual embryo culture. Reprod. Fertil. Dev. **22**, 32 (2010)
21. Lai, D., Smith, G.D., Takayama, S.: Lab-on-a-chip biophotonics: its application to assisted reproductive technologies. J. Biophotonics. **5**, 650–660 (2012)
22. Hwang, H., Lu, H.: Microfluidic tools for developmental studies of small model organisms-nematodes, fruit flies, and zebrafish. Biotechnol. J. **8**, 192–205 (2013)
23. Khalilian, M., Navidbakhsh, M., Valojerdi, M.R., Chizari, M., Yazdi, P.E.: Estimating Young's modulus of zona pellucida by micropipette aspiration in combination with theoretical models of ovum. J. R. Soc. Interface **7**, 687–694 (2010)
24. Dean, J.: Biology of mammalian fertilization: the role of the zona pellucida. J. Clin. Invest. **89**, 1055–1059 (1992)
25. Green, D.P.L.: Three-dimensional structure of the zona pellucida. Rev. Reprod. **2**, 147–156 (1997)
26. Liu, X., Fernandes, R., Jurisicova, A., Casper, R.F., Sun, Y.: In situ mechanical characterization of mouse oocytes using a cell holding device. Lab. Chip. **10**, 2154–2161 (2010)
27. Murayama, Y., et al.: Omata Mouse zona pellucida dynamically changes its elasticity during oocyte maturation, fertilization and early embryo development. Human Cell **19**, 119–125 (2006)
28. Murayama, Y., et al.: Elasticity measurement of zona Pellucida using a micro tactile sensor to evaluate embryo quality. J. Mamm. Ova. Res. **25**, 8–16 (2008)
29. Khalilian, M., Navidbakhsh, M., Valojerdi, M.R., Chizari, M., Yazdi, P.E.: Alteration in the mechanical properties of human ovum zona pellucida following fertilization: experimental and analytical studies. Exp. Mech. **51**, 175–182 (2011)
30. Liu, X., Shi, J., Zong, Z., Wan, K.-T., Sun, Y.: Elastic and viscoelastic characterization of mouse oocytes using micropipette indentation. Ann. Biomed. Eng. **40**(10), 2122–2130 (2012)
31. Moeendarbary, E., Valon, L., Fritzsche, M., et al.: The cytoplasm of living cells behaves as a poroelastic material. Nature **12**(3), 253–261 (2013)
32. Liu, X., Kim, K., Zhang, Y., Sun, Y.: NanoNewton force sensing and control in microrobotic cell manipulation. Int. J. Robot. Res. **28**, 1065–1076 (2009)
33. Yanez, L.Z., Han, J., Behr, B.B., Pera, R.A.R., Camarillo, D.B.: Human oocyte developmental potential is predicted by mechanical properties within hours after fertilization. Nature Commun. **7**, 10809 (2016)
34. Fauci, L.J., Dillon, R.: Biofluidmechanics of reproduction. Annu. Rev. Fluid Mech. **38**, 371–394 (2006)
35. Osada, H., Tsunoda, I., Matsuura, M., Satoh, K., Kanayama, K., Nakayama, Y.: Investigation of ovum transport in the oviduct: the dynamics of oviductal fluids in domestic rabbits. J. Int. Med. Res. **27**, 176–180 (1999)
36. Anand, S., Guha, S.K.: Mechanics of transport of the ovum in the oviduct. Med. Biol. Eng. Comput. **16**, 256–261 (1998)

37. Paltieli, Y., Weichselbaum, A., Hoffman, N., et al.: Laser scattering instrument for real time in-vivo measurement of ciliary activity in human fallopian tubes. Hum. Reprod. **10**, 1638–1641 (1995)
38. Misra, J.C.: Biomathematics: Modelling and Simulation. World Scientific (2006)
39. Leng, Z., Moore, D.E., Mueller, B.A., Critchlow, C.W., Patton, D.L., et al.: Characterization of ciliary activity in distal Fallopian tube biopsies of women with obstructive tubal infertility. Hum. Reprod. **11**, 3121–3127 (1998)
40. Montenegro-Johnson, Th.D., Smith, A.A., Smith, D.J., Loghin, D., Blake, J.R.: Modelling the fluid mechanics of cilia and flagella in reproduction and development. Eur. Phys. J. E **35**, 111 (2012)
41. Isachenko, V.: In-vitro culture of human embryos with mechanical micro-vibration increases implantation rates. Reprod. BioMed. Online **22**, 536–544 (2011)
42. Ezzati, M., Djahanbakhch, O., Arian, S., Carr, B.R.: Tubal transport of gametes and embryos: a review of physiology and pathophysiology. J. Assist. Reprod. Genet. **31**, 1337–1347 (2014)
43. Eytan, O., Elad, D.: Analysis of intra-uterine fluid motion induced by uterine contractions. Bull. Math. Biol. **61**(2), 221–238 (1999)
44. Li, M.J., Brasseur, J.G.: Nonsteady peristaltic transport in finite-length tubes. J. Fluid Mech. **248**, 129–151 (1993)
45. Mernone, A.V., Mazumdar, J.N., Lucas, S.K.: A mathematical study of peristaltic transport of a Casson fluid. Math. Comp. Model. **35**, 895–912 (2002)
46. Eytan, O., Jaffa, A.L., Elad, D.: Peristaltic flow in a tapered channel: application to embryo transport within the uterine cavity. Med. Eng. Phys. **23**, 473–482 (2001)
47. Blake, J.R., Vann, P.G., Winet, H.: A model of ovum transport. J. Theor. Biol. **102**, 145–166 (1983)
48. Verdugo, P., Lee, W., Halbert, S., Blandau, R., Tam, P.: A stochastic model for oviductal egg transport. Biophys. J. **29**, 257–270 (1980)
49. Aranda, V., Cortez, R., Fauci, L.: Stokesian peristaltic pumping in a three-dimensional tube with a phase-shifted asymmetry. Phys. Fluids. **23**, 081901–081910 (2011)
50. Aranda, V., Cortez, R., Fauci, L.: A model of Stokesian peristalsis and vesicle transport in a three-dimensional closed cavity. J. Biomech. **48**, 1631–1638 (2015)
51. Levi Setti, P,.E., Albani, E., Cavagna, M., Bulletti, C., Colombo, G.V., Negri, L.: The impact of embryo transfer on implantation - a review. Placenta **24**(Suppl.B), S20–26 (2003)
52. Sadeghi, M.R.: The 40th anniversary of IVF: has art's success reached its peak? J. Reprod. Infertil. **19**(2), 67–68 (2018)
53. Lauko, I.G., Rinaudo, P., Dashev S.: A computational parameter study of embryo transfer. Ann. Biomed. Eng. **35**(4), 659–671 (2007)
54. van Weering, H.G., Schats, R., McDonnell, J., Vink, J.M., Vermeiden, J.P., Hompes, P.G.: The impact of the embryo transfer catheter on the pregnancy rate in IVF. Hum. Reprod. **17**, 666–670 (2002)
55. Eytan, O., Zaretsky, U., Jaffa, A.J., Elad, D.: In vitro simulations of embryo transfer in a laboratory model of the uterus. J. Biomech. **40**, 1073–1080 (2007)
56. Lambers, M.J., Dogan, E., Kostelijk, H., Lens, J.W., Schats, R., Hompes, P.G.: Ultrasonographicguided embryo transfer does not enhance pregnancy rates compared with embryo transfer based on previous uterine length measurement. Fertil. Steril. **86**, 867–872 (2006)
57. Mansour, R.T., Aboulghar, M.A.: Optimizing the embryo transfer technique. Hum. Reprod. **17**, 1149–1153 (2002)
58. Frankfurter, D., Trimarchi, J.B., Silva, C.P., Keefe, D.L.: Middle to lower uterine segment embryo transfer improves implantation and pregnancy rates compared with fundal embryo transfer. Fertil. Steril. **81**, 1273–1277 (2004)

59. Tiras, B., Polat, M., Korucuoglu, U., Zeyneloglu, H.B., Yarali, H.: Impact of embryo replacement depth on in vitro fertilization and embryo transfer outcomes. Fertil. Steril. **94**, 1341–1345 (2010)
60. Rovei, V., et al.: IVF outcome is optimized when embryos are replaced between 5 and 15 mm from the fundal endometrial surface: a prospectivf Yanez e analysis on 1184 IVF cycles. Reprod. Biol. Endocrinol. **11**, 114 (2013)
61. Cenksoy, P.O., Fıcıcıoglu, C., Yesiladali, M., Akcin, O.A., Kaspar, C.: The importance of the length of uterine cavity, the position of the tip of the inner catheter and the distance between the fundal endometrial surface and the air bubbles as determinants of the pregnancy rate in IVF cycles. Eur. J. Obstet. Gynecol. Reprod. Biol. **172**, 46–50 (2014)
62. Caanen, M.R., et al.: Hompes embryo transfer with controlled injection speed to increase pregnancy rates: a randomized controlled trial. Gynecol. Obstet. Invest. **81**, 394–404 (2016)
63. Groeneveld, E., et al.: Standardization of catheter load speed during embryo transfer: comparison of manual and pump-regulated embryo transfer. Reprod. Biomed. Online **24**(2), 163–169 (2012)
64. Schoolcraft, W.B.: Importance of embryo transfer technique in maximizing assisted reproductive outcomes. Fertil. Steril. **105**(4), 855–860 (2016)
65. Bilalis, D., et al.: Use of different loading techniques for embryo transfer increasing the risk of ectopic pregnancy. Hum. Reprod. (Abstract Book, P-479) **17**, 162 (2002)
66. Yaniv, S., Elad, D., Jaffa, A.J., Eytan, O.: Biofluid aspects of embryo transfer. Ann. Biomed. Eng. **31**, 1255–1262 (2003)
67. Feng, J.Q.: A long gas bubble moving in a tube with flowing liquid. Int. J. Multiph. Flow **35**, 738–746 (2009)
68. Leverett, L.B., et al.: Red blood cell damage by shear stress. Biophys. J. **12**(3), 257–273 (1972)
69. Parfitt, H.S., Davies, S.V., Tighe, P., Ewings, P.: Red cell damage after pumping by two infusion control devices (Arcomed VP 7000 and IVAC 572). Transfus. Med. **17**(4), 290–295 (2007)
70. Wilson, A.M.M.M., Peterlini, M.A.S., Pedreira, M.L.G.: Infusion pumps and red blood cell damage in transfusion therapy: an integrative revision of the academic literature. Rev. Latino-Am. Enfermagem. **24**, e2763 (2016). https://doi.org/10.1590/1518-8345.1155.2763
71. Goubergrits, L., Affeld, K.: Numerical estimation of blood damage in artificial organs. Artif. Organs **28**(5), 499–507 (2004)
72. Yeleswarapu, K.K., et al.: A mathematical model for shear-induced hemolysis. Artif. Organs **19**(7), 576–582 (1995)
73. Xie, Y., Wang, F., Zhong, W., et al.: Shear stress induces preimplantation embryo death that is delayed by the zona pellucida and associated with stress-activated protein kinase mediated apoptosis. Biol. Reprod. **75**, 45–55 (2006)
74. Kizilova, N., Batyuk, L., Baranets, V.: Human red blood cell properties and sedimentation rate: a biomechanical study. In: Arkusz, K., Będziński, R., Klekiel, T., Piszczatowski, S. (eds.) BIOMECHANICS 2018. AISC, vol. 831, pp. 3–22. Springer, Cham (2019). https://doi.org/10.1007/978-3-319-97286-2_1
75. Elad, D., Jaffa, A.J., Grisaru, D.: Biomechanics of early life in the female reproductive tract. Physiology **35**, 134–143 (2020)
76. Xi, Q., Yang, Q., Wang, M., et al.: Individualized embryo selection strategy developed by stacking machine learning model for better in vitro fertilization outcomes: an application study. Reprod. Biol. Endocrinol. **19**(1), 53 (2021)

Mechanical Investigation for the Use of Polylactic Acid in Total Hip Arthroplasty Using FEM Analysis

Emre Celik[1], Furkan Alemdar[1], Murat Bati[1], Muhammed Furkan Dasdemir[1], Onur Alp Buyukbayraktar[1], K. N. Chethan[2], Mustafa Kara[1], and Şenay Mihçin[1](✉)

[1] Department of Mechanical Engineering, Izmir Institute of Technology, Urla, Izmir, Turkey
senaymihcin@iyte.edu.tr

[2] Department of Aeronautical and Automobile Engineering, Manipal Institute of Technology, Manipal Academy of Higher Education, Manipal, Karnataka 576104, India

Abstract. Polylactic acid (PLA) is a biodegradable non-toxic, biocompatible polymer used as a popular filament material in biomedical applications with the advance of 3D printing technologies. PLA is considered a suitable implant material due to its contribution to bone regeneration. In this study, the use of PLA in Total Hip Arthroplasty (THA) as a liner material was assessed. In this regard, the PLA liner with different material combinations in THA was examined to provide evidence for its potential. The hip implant prototypewas drawn using a computer-aided design tool then transferred into a commercial finite element analysis (FEA) software. The prototypesconsisted of assemblies of PLA with titanium, chrome cobalt, stainless steels, dense NiTi shape-alloys, and Alumina-Zirconia. Simulations were run under static loading conditions. To evaluate and compare the results for the optimum design; factor of safety, total deformation and von Mises stress analysis were used. The results show that Co-Cr implemented implants produce the highest factor of safety. When Al-Zi combined with PLA, it produced least deformation and reasonable von- Mises stress values. PLA might perform best when used with Al-Zi. As a next step, experimental pre-clinical tests are planned to assess the clinical potential.

Keywords: PLA · FEA · Total Hip Arthroplasty · Plastic liner · Biomaterial

1 Introduction

From the past to the present, the need for hip implants has been increasing, so the search for longer-lasting and cheaper materials has occurred. UHMWPE is the most used material as a liner material but in this study, Polylactic acid (PLA) is used as an alternative to this material and a comparison is made. PLA is one of these materials which is thought to have the capability to meet these requirements. PLA is a biodegradable material derived from renewable sources such as corn starch or sugar-containing plants. It is produced by polymerization method using lactic acid and lactide monomers [1]. Polylactic acid (PLA) has been a popular material recently due to the advance of 3D printing technologies [2].

A. Hadamus et al. (Eds.): BIOMECHANICS 2021, LNNS 328, pp. 17–23, 2022.
https://doi.org/10.1007/978-3-030-86297-8_2

PLA is considered as a promising material in the biomechanics field due to its biocompatible, osteo-integrative, biodegradable properties [3]. In this research, the potential of PLA use as a liner in Total Hip Arthroplasty (THA) is investigated. The plastic liner is used to reduce friction and wear in hip implants and placed between the cup and head parts [4].

One of the most important factors affecting material selection is its compatibility with surrounding tissues. Care has been taken to ensure that the selected materials are biocompatible. In this study, PLA material and UHMWPE, which is frequently used material as a liner, were compared. In this study, alternative material properties were investigated by mechanical model. Changes in results were observed by using different materials under the same mechanical properties.

One of the important issues in material selection is Young's Modulus. However, in mechanical calculations, not only the material, but also the total Young's Modulus should be considered. Changing the mechanical state causes bone adaptation and decreased strength because of resorption of excess tissue. In the present formulation, the timescale of tissue evolution is much longer than is required for analysis of the mechanical transient effect. Therefore, the evolution of Young's modulus is considered a quasi-static processs [5]. The osteophytes in growth is a mechanical and biological disease that occurs in a region close to the cartilage surface and in the bone tissue connection. Although there is not much research on the subject, mechanical aspects are very important in this disease [6]. The selection of materials compatible with bone has gained great importance in the last few decades. Traditionally, bioinert materials such as stainless steel, titanium and its alloys have been used. However, when such materials are used, a second surgery is usually needed. However, due to the high hardness of metallic materials, there is a risk of unwanted bone resorption caused by stress shielding [7]. In implant design, it is very important to determine the material properties of the tissue such as elastic modulus, hardness, and its ability to harden during the deformation process. The most important factor is the chosen material. In many studies, tests have been made on tissue samples and material selection has been carried out by using the results of these tests [8]. Some models, algorithms and approaches have been used to describe bone tissue mechanics and remodeling processes in hip implant design [9].

In this study we focus on computational modelling of the hip implants. Our study is limited to numerical analysis and experimental work could be performed as a next step to investigate the interaction of the implant with the surrounding components. Here, we used the classical Cauchy continuum deformable body model for isotropic materials in this study.The main objective of this study is to investigate the mechanical behaviour of PLA material in THA assembly using Finite Element Modelling (FEM). The results are expected to provide evidence about the mechanical behaviour of PLA in THA.

2 Materials and Methods

The geometry of the models was defined in SolidWorks and drawn in ANSYS software and transferred into ANSYS R-19 (ANSS, Canonsburg, Pennsylvania, USA).The model used for the samples is the classical Cauchy continuum deformable body for isotropic materials. The stem part of the model was designed by the authors considering the examples taken from the literature. According to the literature, it may vary according to the implant geometry of the fixed support. Considering the selected fixed supports of similar geometry types in the literature, it was deemed appropriate to choose half of the stem. Contacts and their coefficient of friction were examined. Our first contact is between the cup and liner parts. The friction contact, and its coefficient is defined around 0.2. The second contact is between the liner and parts of the femoral head. The contact's name is friction contact, and its coefficient is again 0.2. And the final contact is between the femoral head and body parts. Here, two faces are chosen so that the head does not slip. Designed element size is selected as 2 mm in all parts to be able to get more accurate results. In FE model, the mesh has 55012 elements which is very high and time-consuming however this is necessary to obtain accurate results. The used mesh type is face sizing. Five different models were created for five different materials. In these prototypes, the liner material is PLA, and all other parts of the implants are made of the same materials because of the investigations in this field. These materials are Co-Cr, Stainless Steel, Ti6Al7Nb, Al-Zi, Dense Niti Shape Alloys [10]. Material properties were used from the ANSYS material library. In this study, Young's modulus, Poisson's ratio and mass density values for all materials were obtained from Ansys database. The force is applied as a surface force on the cup. The components are F_x equals to 0 N, F_y equals to 1410 N, F_z equal to −1820 N in a general coordinate system (Fig. 1) in ANSYS [11]. In this study, friction coefficients were assumed as approximately 0.2.Although these are not exact values for coefficients of friction, the exact values are all close to 0.2, hence this approximation was made [12–16]. The safety factor is calculated by dividing the strength of the material by the maximum stress.

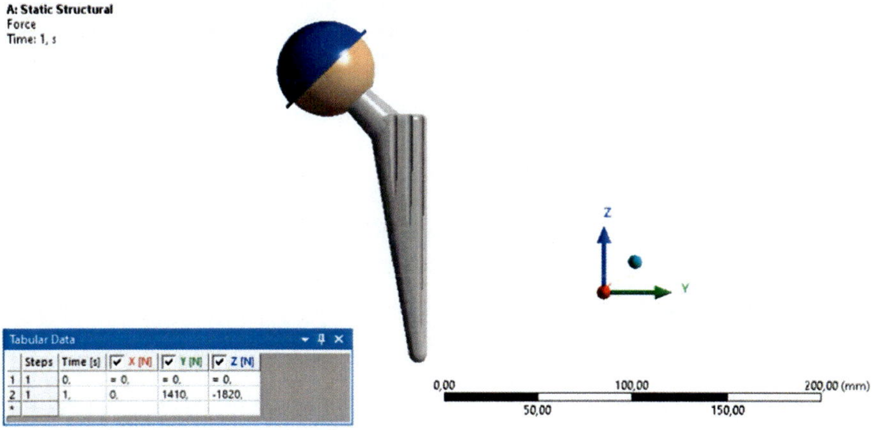

Fig. 1. Model indicating the coordinate system and defined force values.

3 Results

Our simulation shows that PLA material exhibits the best resistance to loading, friction forces when combined with the Al-Zi. The calculated safety factor for Al-Zi is 4.6, the total deformation is 0.021 mm and the von Mises stress value is 103.23 MPa under the specified boundary conditions mentioned above. To test the new combinations in more detail, more analyses were performed with varying boundary conditions. As shown in Table 1, Dense-NiTi Shape Alloys/PLA and Al-Zi/PLA pairs had the highest values of the von Mises stress. Conversely, Ti6Al7Nb/PLA showed the lowest von Mises stress value. Regarding total deformation, Dense-NiTi Shape Alloys/PLAandAl-Zi/PLA had the maximum and minimum values respectively amongst all examined groups. Moreover, the highest and lowest values of factor of safety corresponded to Co-Cr/PLA and SS314/PLA respectively. Figure 2 also indicates the total deformation contours of all models. According to this figure, the maximum. deformation occurred in the shell component. The maximum deformation values were clearly labeled in the figure below.

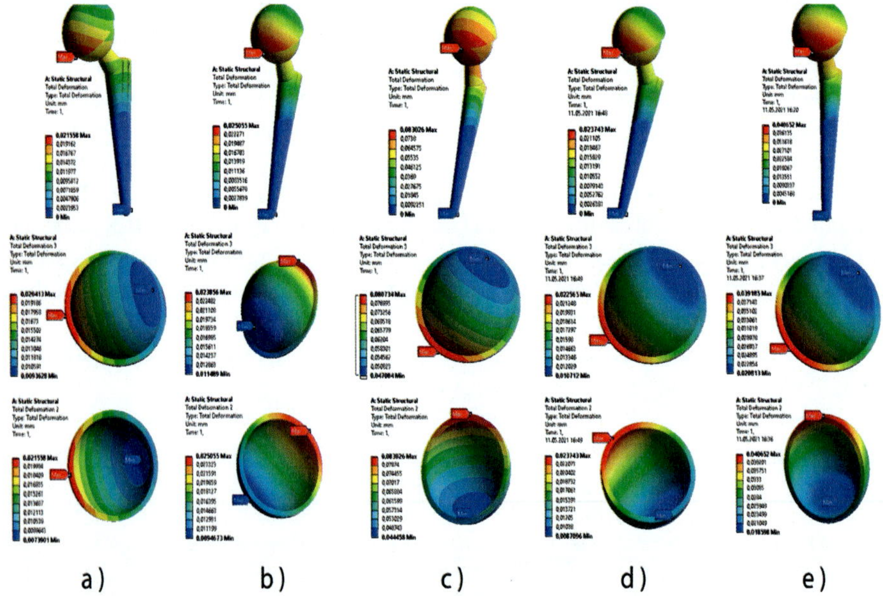

Fig. 2. Total Deformation Analysis: (a) Alimuna-Zirconia (b) Co-Cr (c) Dense Niti Shape Alloy (d) SS314 (e) Ti6AlNb7.

Table 1. Results for Hip implant with PLA and UHMWPE Liner under static loading.

Material	Total deformation (mm)	Von Mises stress (MPa)	Safety factor
Co-Cr // PLA	0.025	96.92	9.52
Ti6Al7Nb // PLA	0.041	91.41	6.11
Dense-NiTi Shape Alloys // PLA	0.083	103.23	7.66
SS314 // PLA	0.024	99.12	2.19
Al-Zi // PLA	0.021	103.23	4.63
Co-Cr // **UHMWPE**	0.044	96.92	9.52
Ti6Al7Nb // **UHMWPE**	0.059	91.41	6.11
Dense-NiTi Shape Alloys//**UHMWPE**	0.101	103.23	7.66
SS314 // **UHMWPE**	0.043	99.12	2.19
Al-Zi **// UHMWPE**	0.041	103.23	4.63

4 Discussions

Currently in THA, UHMWPE is used widely as a liner material. According to the analysis performed so far on the liner made of both UHMWPE and PLA, it was observed that both of the liner materials gave similar results under same analysis modelling conditions. The friction coefficients differ for each material. These values are around the 0.2 value. For this reason, it was accepted as 0.2. In terms of von Mises stress and safety factor values, both two material showed completely same results. Different liner materials affect only the total deformation values. In both analysis, Al-Zi showed the best total deformation results. The total deformation of Al-Zi with UHMWPE liner is 0.041mm and with PLA, it is 0.021mm. According to these analyses, PLA gave smaller total deformation values than UHMWPE and it could be used instead of UHMWPE. According to the literature, oval shape profile implants mostly have 730 MPa von Mises stress values [17]. That comparison shows us that PLA may be a new future material for hip implant liner.

However, the above conclusion is solely based on computational modelling results after running the simulation *in silico*. In literature, there are many other studies [5–9] focusing on bone interaction, osseo integration, dealing with tissue mechanics. The study could be further developed considering the bone and surrounding tissues, to investigate for practical applications to provide more evidence before any animal studies take place with the proposed materials.

5 Conclusion

The results show that material combinations are of great importance while deciding on a THA assembly. PLA performs best when used it Al-Zi. PLA could be an exciting material for medical applications with its promised high corrosion resistance, high abrasion resistance and other specific properties. Although PLA material has promising results, wear tests should also be performed to investigate the wear properties, supported by the experimental studies next in preclinical settings. Nevertheless, clinical implications need to be investigated and there is a need for experimental data for the reliability of PLA material use. This study forms the baseline for the fundamental research to lead the way to possible combined applications of PLA in the future.

Acknowledgments. The authors thank the Mechanical Engineering, Department of Izmir Institute of Technology for providing the high computational facility to carry out this research.

Funding. This research was funded by TUBITAK 2232 International Outstanding Researchers Funding Scheme with Grant No of 118C188 'New Generation Implants for All' project.

Declarations. Ethics Approval and Consent to Participate
Not applicable.

Consent for Publication. All authors consent to the publication of this manuscript.

Competing Interests. The authors declare that they have no competing interests.

References

1. Tomčíková, Z., Ujhelyiová, P., Krivoš, M.Š., Hricová, M.: Influence of plasticizer and bioplasticizer on the structure and mechanical properties of the plafibres. Vlakna A Textil **24**(4), 51–57 (2017)
2. Okolie, O., Stachurek, I., Kandasubramanian, B., Njuguna, J.: 3D printing for hip implant applications: a review. Polymers **12**(11), 2682 (2020)
3. Dedukh, N.V., Makarov, V.B., Pavlov, A.D.: Polylactide-based biomaterial and its use as bone implants (analytical literature review). Pain Joints Spine **9**(1), 28–35 (2019)
4. Massimiliano, M., Affatato, S.: Materials for hip prostheses: a review of wear and loading considerations. Materials **12**(3), 495 (2019)
5. Lu, Y., Lekszycki, T.: A novel coupled system of non-local integro-differential equations modelling Young's modulus evolution, nutrients' supply and consumption during bone fracture healing. Z. Angew. Math. Phys. **67**(5), 1–12 (2016). https://doi.org/10.1007/s00033-016-0708-1
6. Bednarczyk, E., Lekszycki, T.: A novel mathematical model for growth of capillaries and nutrient supply with application to prediction of osteophyte onset. Z. Angew. Math. Phys. **67**(4), 1–14 (2016). https://doi.org/10.1007/s00033-016-0687-2
7. Lu, Y., Lekszycki, T.: New description of gradual substitution of graft by bone tissue including biomechanical and structural effects, nutrients supply and consumption. Continuum Mech. Thermodyn. **30**(5), 995–1009 (2018). https://doi.org/10.1007/s00161-018-0650-x

8. Pawlikowski, M., Skalski, K., Bańczerowski, J., Makuch, A., Jankowski, K.: Stress–strain characteristic of human trabecular bone based on depth sensing indentation measurements. Biocybern. Biomed. Eng. **37**(2), 272–280 (2017)
9. Giorgio, I., Spagnuolo, M., Andreaus, U., Scerrato, D., Bersani, A.M.: In-depth gaze at the astonishing mechanical behavior of bone: a review for designing bio-inspired hierarchical metamaterials. Math. Mech. Solids **26**(7), 1074–1103 (2021)
10. Hu, C.Y., Yoon, T.-R.: Recent updates for biomaterials used in total hip arthroplasty. Biomater. Res. **22**(1), 1–12 (2018)
11. Chethan, K.N., et al.: Static structural analysis of different stem designs used in total hip arthroplasty using finite element method. Heliyon **5**(6), E01767 (2019)
12. Chowdhury, M.A., et al.: Experimental investigation on friction and wear of stainless steel 304 sliding against different pin materials. World Appl. Sci. J. **22**(12), 1702–1710 (2013)
13. Guezmil, M., Bensalah, W., Mezlini, S.: Tribological behavior of UHMWPE against TiAL6V4 and CoCr28mo alloys under dry and lubricated conditions. J. Mech. Behav. Biomed. Mater. **63**, 375–385 (2016)
14. Perepelkina, S., et al.: Investigation of friction coefficient of various polymers used in rapid prototyping technologies with different settings of 3D printing. Tribol. Ind. **39**(4) (2017)
15. Triyono, J., et al.: Investigation of meshing strategy on mechanical behaviour of hip stem implant design using FEA. Open Eng. **10**(1), 769–775 (2020)
16. Bahraminasab, M., Sahari, B.B.: NiTi shape memory alloys, promising materials in orthopedic applications. shape memory alloys-processing. Character. Appl., 261–278 (2013)
17. Yongtae, J., Kuiwoon, Ch.: Design of Patient-Specific Hip Implants Based on the 3D Geometry of the Human Femur. Elsevier Ltd., Korea (2009)

Numerical Determination of the Degree of Mechanical Anisotropy of the Femoral Neck Trabecular Bone

Artur Cichański(✉) and Krzysztof Nowicki

UTP University of Science and Technology in Bydgoszcz, Al. prof. S. Kaliskiego 7, 85-796 Bydgoszcz, Poland
artur.cichanski@utp.edu.pl

Abstract. The paper deals with the problem of determining of parameters describing the degree of trabecular bone structure anisotropy. The Mean Intercept Length (MIL), the most commonly used characteristics of anisotropy in trabecular bone, was taken as a gold standard method. As an alternative description, an approach was proposed in which the degree of anisotropy was defined as the ratio of the apparent modulus determined by mFE in two orthogonal directions. For the analyses, an important issue was to determine the direction of occurrence of the maximum value of the apparent modulus. The paper also presents the procedure of calibrating the numerical model in terms of the correct definition of the boundary conditions for analyses carried out with trabecular tissue samples cut at rotated coordinate system.

Keywords: Cancellous tissue · Apparent stiffness · mFE analysis · MIL method

1 Introduction

Predicting the strength of a cancellous bone is interesting for its ability to diagnose bone fracture risk and better understand the mechanisms that influence bone strength. Experimental studies in this area confirm the existence of a relationship between bone strength and its stiffness [1]. The stiffness of trabecular structures is also determined in the course of micro-finite element (mFE) simulations using models created on the basis of microtomograph images (mCT) [2]. For mFE models, the size of the voxel and the method of its defining have a significant impact on the determined value of trabecular tissue stiffness [3].

Trabecular bone is a tissue that undergoes constant remodelling that depends on the equilibrium between the processes of bone cells formation and bone cells resorption [4]. This process occurs in response to mechanical stimuli and is associated with the bone morphology [5]. Since mechanical stimuli differs from site to site of the body, the remodelling process is site dependent as well and the bone resorption and formation are strongly correlated with local tissue loading [6]. Pathological changes in the bone remodelling process leads do the rapid loss of trabeculae, that is the main reason of osteoporosis. If bone cells become less mechanosensitive the possibility of the rapid loss

A. Hadamus et al. (Eds.): BIOMECHANICS 2021, LNNS 328, pp. 24–36, 2022.
https://doi.org/10.1007/978-3-030-86297-8_3

of bone mass increases [7]. This remodelling process generates anisotropy in the trabecular bone. To characterise that anisotropy is used fabric tensor [8]. Elastic properties of the cortical bone at the macroscale may be described with assumption of orthotropic symmetry [9]. The mCT images could be used to estimate the degree and the directions of orthotropy [10]. The other microstructure feature influences bone remodelling is porosity. The decrease in porosity cause a proportional decrease in the mechanical stimuli that affects of bone regeneration [5]. Trabecular bone can be characterized as a poro-viscoelastic material due to the bone tissue intrinsic viscoelasticity [11].

The stiffness of the trabecular bone is influenced not only by the content of the mineral fraction but also by its distribution. Principal Component Analysis of the relationship between bone structure indices and mechanical properties indicates that one of the more independent indices is the degree of anisotropy (DA) [12] that is determined by Mean Intercept Length (MIL) method. Mean intercept length represents the mean distance between intersections of bone and non bone components with set of parallel lines designated in a particular orientation of measurements. The spatial distribution of tissue described by this method is approximated by an ellipsoid. The degree of anisotropy DA is defined as ratio between the maximum (MIL1) and minimum (MIL3) mean intercept length, which are measured at the major (1) and minor (3) axis of ellipsoid. The accuracy of DA determination does not depend on either the scanning resolution or the resolution of the structure reconstruction [13]. In order to ensure the representativeness of the apparent module determined mFE, models are processed along it principal mean intercept length MIL axis [14]. Correct determination of the structure directivity in the MIL analysis requires that the sample used for calculations must be at least 1 mm high in the axial direction [15]. The structural anisotropy is described by bone structure indicator DA. To the description of mechanical anisotropy the DM parameter may be used [16]. Based on results of experimental compression of cubic specimens from human femoral heads the degree of mechanical anisotropy DM was defined as the ratio between modulus determined in two orthogonal directions. The experimentally determined values of DM are about 50% higher than the DA values for trabecular bones [16]. The DM parameter for cortical bone is lower [17] than for cancellous bone tissue [16].

The aim of the study was to compare the effectiveness of MIL and mFE analysis to determine the degree of anisotropy for the femoral neck bone tissue structure. While the direction of occurrence of the maximum stiffness for the MIL was directly determined, an important step in achieving the goal was to indicate this direction during the mFE analyzes. The correct performance of mFE calculations required the development of a method of defining boundary conditions.

2 Materials and Methods

2.1 Materials

Tests were performed on samples of the trabecular bone structure. Specimens were obtained from 33 patients by dissecting them from the femoral neck obtained after hip arthroplasty surgery. After the collection, the slices of the structure were stored in a way that did not change their mechanical properties [18]. Cylindrical samples with a diameter of ø10 mm and a height of 7.7 mm were cut from the slices. Samples were scanned with a resolution of 36 μm on a microtomograph μCT80 (SCANCO Medical AG, Switzerland). The verification of the proposed method was performed on the basis of 11 samples selected from set n = 33 based on bone volume fraction (BV/TV). Structural indices of the selected samples such as trabecular number (Tb.N), trabecular thickness (Tb.Th) and connectivity density (Conn.D) are presented in Table 1.

Table 1. Structural indices of the selected samples.

Decyle D of BV/TV	BV/TV, -	Tb.N, 1/mm	Tb.Th, mm	Conn.D
Min	0.151	1.138	0.1329	3.097
1th	0.168	1.018	0.1656	2.151
2nd	0.175	1.144	0.153	2.969
3th	0.185	1.241	0.1492	3.611
4th	0.193	1.105	0.1751	2.565
5th	0.201	1.086	0.1853	2.158
6th	0.234	1.415	0.1653	3.959
7th	0.248	1.305	0.1899	3.034
8th	0.272	1.408	0.1933	3.290
9th	0.298	1.271	0.2345	2.579
Max	0.353	1.497	0.236	3.980

2.2 Methods of Modelling

As a result of the scanning, 214 images were obtained for each sample. Based on the images, geometric models were prepared for cylindrical samples. From the images with a resolution of 36 μm, models were created for which the voxel size was a multiple of 1×, 2×, 3× of the scan resolution. In this way, cylindrical samples models with voxel sizes d = 36 μm, 72 μm and 108 μm were prepared. The method of bone structure reconstruction has been presented as a 3D method in [3]. Numerical analysis were performed using the APDL component of the ANSYS (Release 18.2, ANSYS, Inc.). The isotropic material properties $E = 10$ GPa and $\nu = 0.3$ were adopted for the analyzes [19].

Based on geometric models of cylindrical samples a FE mesh were prepared by conversion voxels to hexahedral elements with a side of 36 μm. The cube-shaped samples with a side of 5.4 mm were cut from FE models of the cylindrical samples with a diameter of ø10 mm (Fig. 1). The center of the cut cube lay in the middle of cylinder's height on its symmetry axis. Successive cubes were cut by changing the angles α measured in the 0XY plane and the angle γ measured in the 0XZ plane (Fig. 1). Disconnected elements were removed after cutting the cubic sample from cylindrical sample. Slice of sample BV/TV = 0.248 limited to elements lying in the first octant of the spatial coordinate system shown in Fig. 2.

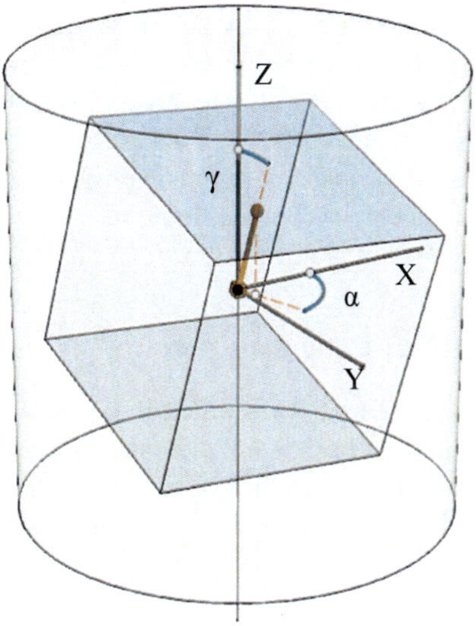

Fig. 1. Scheme for defining sample cutting angles.

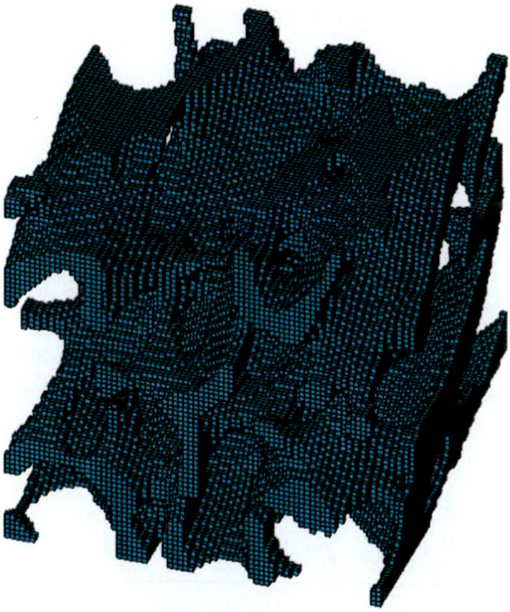

Fig. 2. Trabecular structure sample slice, $d = 36$ μm, BV/TV = 0.248, $\alpha = 0°$, $\gamma = 0°$.

The boundary conditions for the analyzes caused compression of the cubic sample in the direction of the 0Z axis by applying appropriate displacements to the nodes located near the shaded faces of the cube Fig. 1. The nodes located near the shaded sample face laying in negative orientation of OZ axis were fixed in the OZ direction. To the nodes located near the shaded sample face laying in positive orientation of OZ axis such displacement was applied to achieve compression $\varepsilon = 1.0\%$. Based on the results of the analyzes, the apparent Young's modulus for the cubic sample was determined.

2.3 Calibration of Boundary Conditions Zone

For the analyzes, it was assumed that the mesh will be, both inside the sample and on the outer surfaces, will consist of hexahedral elements as they were created by converting voxels built on the basis of threshold images. The geometric consequence of this assumption was the jagged edges of the outer surface of samples whose axes are rotated in relation to the global coordinate system Fig. 3. The extreme nodes of the elements making up these surfaces were laying not in one plane. In order to define the boundary conditions during such analyzes, it was necessary to select nodes lying not only on the sample surface but also inside, near the sample surface. The thickness of this nodes layer t influenced the calculated apparent stiffness of the sample. In order to determine that thickness, analysis were carried out using a homogeneous test model of the perfect porous structure Fig. 3.

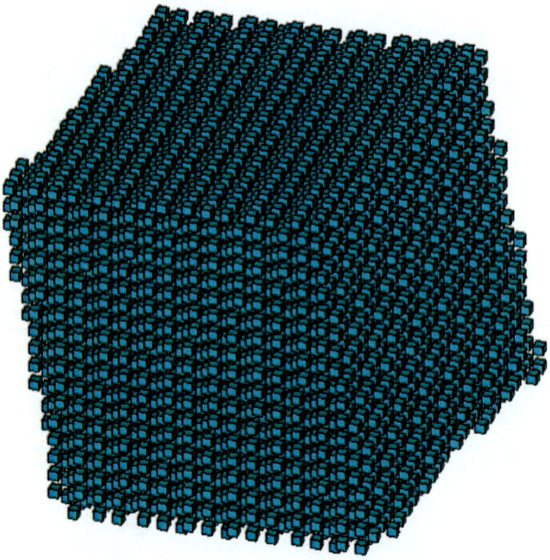

Fig. 3. A slice of perfect porous structure, $d = 72$ µm, BV/TV = 0.25, $\alpha = 10°$, $\gamma = -10°$.

The perfect porous structure was created on the basis of the base cube with a side of 9.35 mm, which corresponded to the diameter of the sphere described on the cube sample with side 5.4 mm accepted for analysis. Successive elements, 36 µm in size, were moved apart by 36 µm in the directions of the three axes of the Cartesian coordinate system. The regular porous structure thus formed was filled with elements in 25% of the volume, which corresponds to BV/TV = 0.25.

For the preliminary computing a constant thickness of the nodes boundary layer t adopted for the boundary conditions definition was assumed. The value of t was determined iterative by comparing the value of the apparent modulus computed in different α-directions with the reference value. As the reference value the apparent modulus determined for a sample of the perfect porous structure cut out according to the axes of the global coordinate system ($\alpha = 0°$, $\gamma = 0°$) was adopted. The surfaces of such a sample were smooth because all the nodes of a given wall layed in one plane. The thickness value t was accepted for the calculations that were performed for multiple angle α values $0° < \alpha < 90°$ with a step of 5° and single value of angle $\gamma = 0°$. For each α-direction the value of apparent modulus was computed. The mean value of the apparent modulus was representative for the set of analysis carried out over the entire range of α angles. If the mean value of the computed apparent modulus was lower than the reference the thickness t was increased, otherwise thickness t was decreased. The entire procedure was repeated until the mean value of the apparent modulus for the set of analysis was equal to the reference modulus with 0.1% tolerance. In the course of subsequent analyzes, the thickness of the nodes boundary layer was determined depending on the size of the voxel $t = 1.8 \times d$.

In the basic computing the samples were rotated by angles $0° < \gamma < 90°$ with a step of 5°. For each position described by the γ angle, a set of analyzes was performed for samples cut at angles $0° < \alpha < 45°$ with a step of 5°. Along with the change of the angles α and γ of sample cutting, the values of the apparent modulus E changed. Both the maximum and minimum values of E in the considered range of variation of the angle α obtained for individual positions of γ during the analyzes for voxel $d = 108$ μm are presented in Fig. 4. The occurrence of different extreme values of the apparent modulus E for individual angles γ indicates that this modulus depends on this angle. Moreover, the apparent modulus E for individual γ angles reached extreme values for other α angles.

Relative errors of apparent modulus computing δE for individual angles γ compiled for the considered sizes of voxels $d = 36$ μm, 72 μm and 108 μm are shown in Fig. 5. As the voxel size increases, the considered relative error δE increases. Moreover, this error changes irregularly with the change of the angle γ, which indicates that the thickness t sought for the boundary layer of nodes adopted for the boundary conditions cannot be a constant value that depend only on voxel size d.

Due to the dependence of the calculation accuracy on both the α and γ angle, the thickness t of the nodes layer adopted for the boundary conditions should depend not only on the voxel size but also on both angles. The search for the function $t = f(d, \alpha, \gamma)$ describing that condition was carried out in two stages. In the first stage, for voxels $d = 36$ μm and angles $\alpha = 2°$ and $\gamma = 5°$, the value of thickness t was iteratively determined, that led to the determination of the apparent modulus equal to the value of the reference modulus. Then the procedure was repeated for angles $0° < \gamma < 90°$ with a step of 5° and angles $0° < \alpha < 44°$ with a step of 2°. In the second stage, the obtained values of the thickness t were approximated by the multiple regression method using the Statistica 13.1 program. Due to the nature of the volatility of the error δE the considered

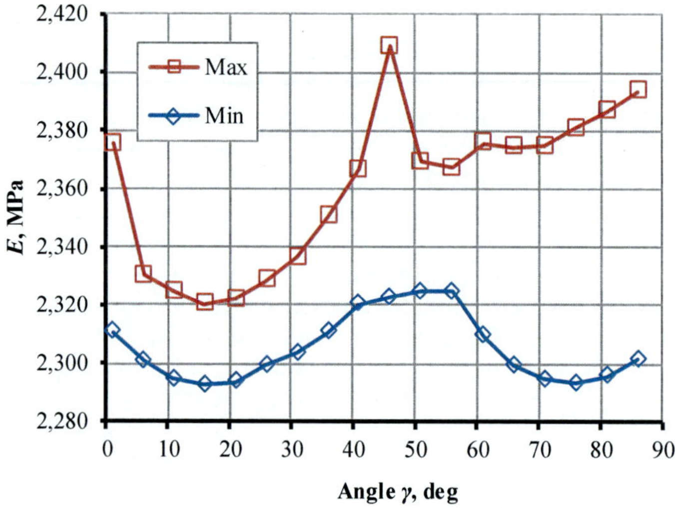

Fig. 4. Extreme values of E, $0° < \alpha < 45°$, $d = 108$ μm.

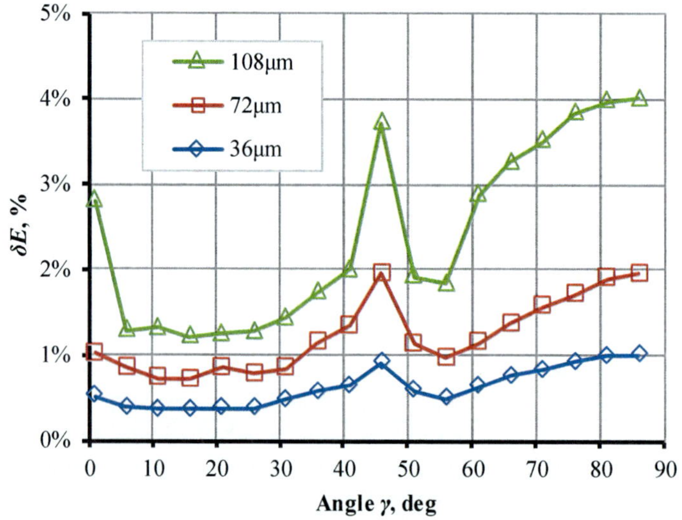

Fig. 5. Relative error δE, thickness of nodes boundary layer $t = \mathrm{f}(d)$.

range of γ angles was divided into three areas and separate approximating function was determined for each area. In the range $0° \leq \gamma < 42°$ the approximating relationship (1) was determined, with the multiple regression coefficient R = 0.9671 and the standard error of estimation 0.0312. In the range $42° \leq \gamma \leq 54°$, the approximating relationship (2) was determined, with the multiple regression coefficient R = 0.9738 and the standard error of estimation 0. 0219. In the range $54° < \gamma \leq 90°$ the approximating relationship was determined (3), with the multiple regression coefficient R = 0.9682 and the standard error of estimation 0.0277.

$$t/d = 1.26235 + 0.03431 \cdot \gamma - 0.00013765 \cdot \alpha \cdot \gamma - 0.001488 \cdot \gamma^2 + 0.0000033 \cdot \alpha \cdot \gamma^2 - 0.0000036 \cdot \alpha^3 + 0.0000157 \cdot \gamma^3 + 0.01344 \cdot \alpha \quad (1)$$

$$t/d = 1.19 + 0.0001236 \cdot \alpha \cdot \gamma + 0.00085 \cdot \alpha^2 + 0.000049 \cdot \gamma^2 - 0.00000837 \cdot \alpha^2 \cdot \gamma - 0.0000112 \cdot \alpha^3 \quad (2)$$

$$t/d = -2.9434 + 0.100205 \cdot \alpha + 0.1207 \cdot \gamma - 0.00195 \cdot \alpha \cdot \gamma - 0.0007865 \cdot \alpha^2 - 0.000812886 \cdot \gamma^2 - 0.0000046 \cdot \alpha^2 \cdot \gamma + 0.000013 \cdot \alpha \cdot \gamma^2 + 0.00001055 \cdot \alpha^3 \quad (3)$$

The obtained approximating relationships were verified using the model of perfect porous structure. Relative errors of apparent modulus δE for considered angle γ are shown in Fig. 6. The maximum error on determining the apparent module for voxels $d = 36$ μm was 0.27%, while for voxels $d = 72$ μm and 108 μm it was less than 1.1%.

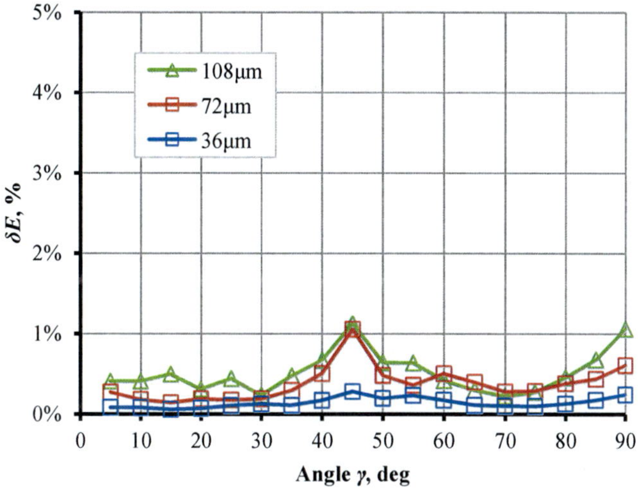

Fig. 6. Relative error δE, thickness of nodes boundary layer $t = f(d, \alpha, \gamma)$.

3 Results

In the course of main numerical analyses, the values of the apparent modulus were determined for samples cut at angles $0° < \gamma < 180°$ with a step of 5°. For each position described by the γ angle, a set of analyzes was performed for samples cut at the angles $0° < \alpha < 360°$ with a step of 5°. The exemplary results of analyses for selected γ angles are presented for the sample BV/TV = 0.298 in Fig. 7.

In order to determine the maximum value of the apparent modulus and the angular coordinates describing it, detailed analyzes were performed for the selected γ angles, narrowing the range of variability of the angle α to the range $0° \div 90°$. The exemplary

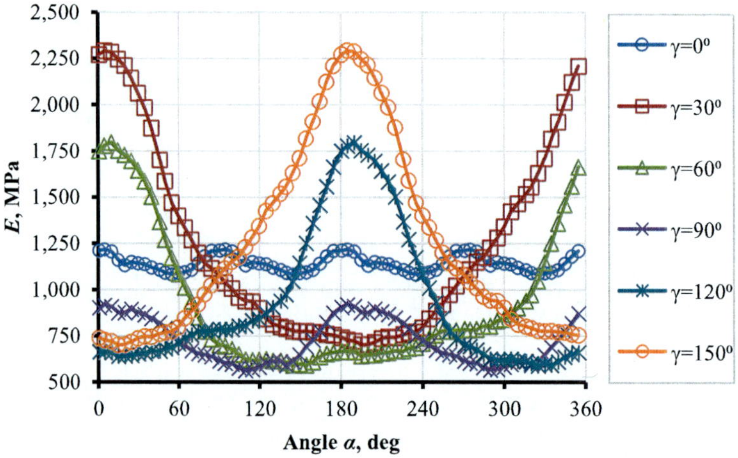

Fig. 7. Apparent modulus $E = f(\alpha)$, sample BV/TV = 0.298, full α range.

results of the analyzes are presented for the sample BV/TV = 0.298 in Fig. 8. On the basis of the so determined courses of variability of the apparent module, it was established that for the sample BV/TV = 0.298 the maximum E is 2466 MPa and occurs for the angle $\gamma = 38°$ and the angle $\alpha = 10°$.

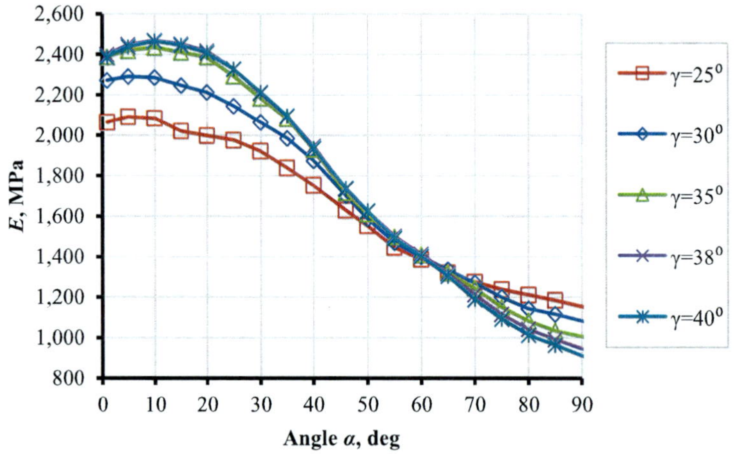

Fig. 8. Apparent modulus $E = f(\alpha)$, sample BV/TV = 0.298, first quarter α range.

Similar analyses were performed for all of the selected samples. Determined by FEM, angular positions $\alpha$and $\gamma$of the occurrence of the apparent modulus maximum and the corresponding them values of modulus are summarized in Table 2. This table also shows the angular positions that describe the position of the MIL1 axis. For these positions, corresponding values of apparent modules were numerically determined.

Table 2. Angular positions for max apparent modules.

BV/TV,	FEM			MIL		
-	α, deg	γ, deg	E, MPa	α, deg	γ, deg	E, MPa
0.151	320	30	459	327	28	432
0.168	350	44	1 200	349	45	1 193
0.175	295	25	1 129	278	13	1 094
0.185	15	38	690	9	43	647
0.193	320	42	1 359	319	40	1 272
0.201	15	24	1 212	79	25	802
0.234	205	45	1 391	220	40	1 170
0.248	260	33	1 915	269	31	1 886
0.272	195	30	1 857	194	28	1 841
0.298	10	38	2 466	12	37	2 451
0.353	295	34	3 078	292	30	3 066

Correlation of the values of γ angles determined by FEM and MIL for the maximum values of the modules presented in Fig. 9a is characterized by the coefficient of determination $R^2 = 0.805$. In Fig. 9b a similar comparison is shown for the angles α. It indicates a high correlation angles of determined by FEM and MIL as well $R^2 = 0.977$. The obtained values of the determination coefficients indicate the existence of a strong relationship between the orientation anisotropy angles of the bone structures determined by both methods. The obtained values of the coefficients of the linear correlation equation indicate the existence of a unit relationship between the angles $\alpha$and γdetermined for both methods. The directional coefficients of the straight lines for the angles γ and α are close to one. On the other hand, the free terms of the equations indicate the existence of a constant error, about 20° for the angle α.and 8° for the angle γ.

The apparent modules for directions perpendicular to the directions of the maximum E were determined. The new orientation of E computing was described by the angles α and $\gamma + 90°$ (Table 2, section FEM). The minimum value of the apparent modulus determined in this way and computed earlier maximum value were used to calculate the degree of mechanical anisotropy of the trabecular structure using the relationship (4).

$$\mathrm{DM} = 1 + \frac{E_{min}}{E_{max}} \tag{4}$$

The values of the degree of structural anisotropy DA of the trabecular structure determined by the MIL analysis and the degree of mechanical anisotropy of the trabecular structure DM determined by the FEM analyses are summarized in Fig. 10. The values of the medians and, respectively, the first and third quartiles of anisotropy degree indices determined for selected samples using both methods are presentedin Fig. 11. For the considered samples, the ratio of mean DA to mean DM was 0.558. A similar ratio

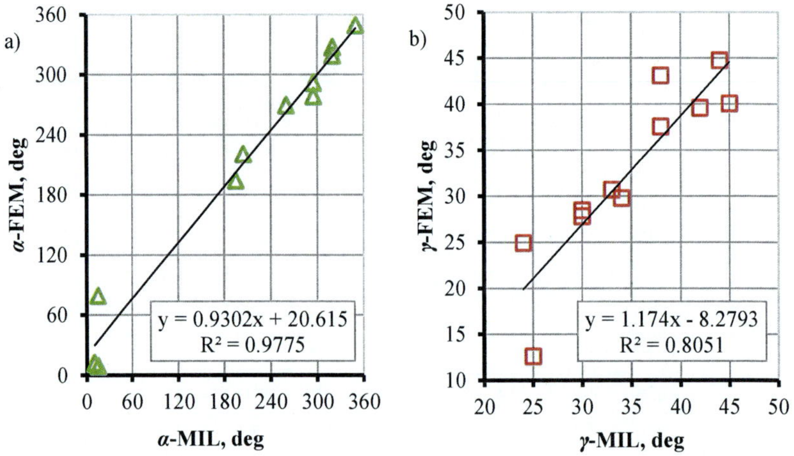

Fig. 9. Comparison of FEM and MIL methods; a) α angles, b) angles γ.

calculated using the experimentally determined DM value was 0.678 for cancellous bone tissue [16].

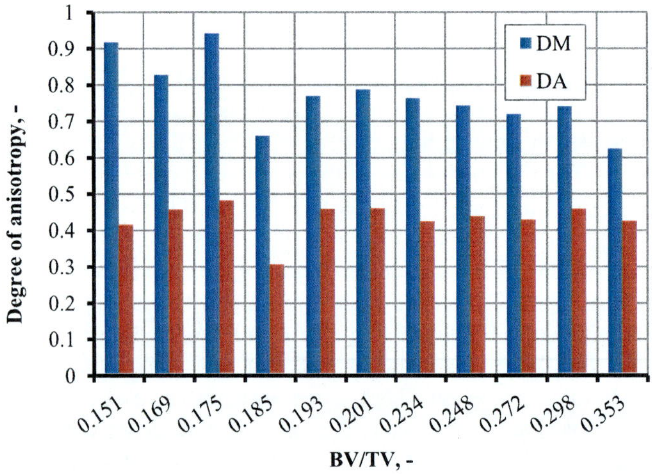

Fig. 10. Comparison of the bone structure parameters DM and DA.

The Shapiro-Wilk tests for the normality of the distribution for the values of DM and DA showed that in the case of DM the distribution can be considered normal (p-value = 0.6239). However, for DA, this hypothesis is invalid (p = 0.0026). Therefore, the non-parametric Wilcox test for the occurrence of differences between the distributions of DM and DA was performed, obtaining the p-value = 8×10^{-5}. Thus, the hypothesis about the compatibility of the distributions may be rejected. By analyzing the graph presented in Fig. 11 it can be concluded that the median value for DM is higher than for DA (difference approx. 0.3). Also, the range of variability is much lower for the DA method than for the DM method (about twice).

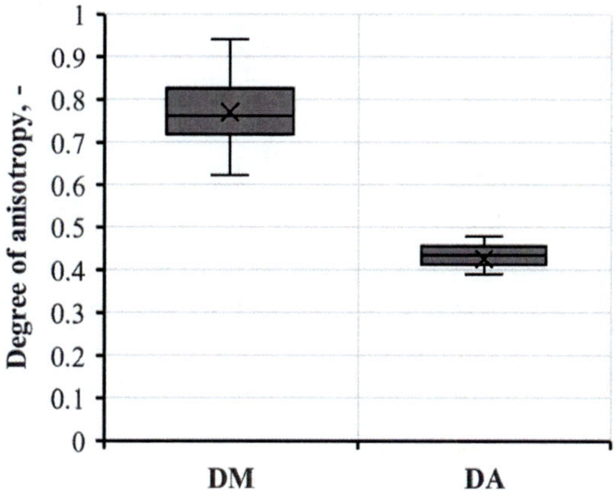

Fig. 11. Comparison of the differences between the bone structure parameters DM and DA.

4 Conclusions

The effectiveness of determining the directions of occurrence of the maximum values of the apparent modulus by the MIL and FEM method is similar, as evidenced by the results presented in Fig. 9.

The values of the DM coefficient are different from the values of the DA coefficient as shown in Fig. 11. Due to the differences in the method of determining the directions of the trabecular bone structure between the methods, it can be indicated that the DM method, that is not based on an ellipsoid approximation as in the case of DA, tends to capture an increase of local tissue stiffness in the structure, that is not the case with the method DA. The use of approximation in the DA method leads to the omission of such trabecular structure features.

The sensitivity of the DM method to local structural disturbances may be proved by a much higher value of the range of variability of the degree of mechanical anisotropy.

From the technical side, it should be noted that the use of mFE to determine the apparent Young's modulus with simultaneous rotation of the sampled structure requires a special approach to defining boundary conditions.

References

1. Yeni, Y.N., Fyhrie, D.P.: Finite element calculated uniaxial apparent stiffness is a consistent predictor of uniaxial apparent strength in human vertebral cancellous bone tested with different boundary conditions. J. Biomech. **34**, 1649–1654 (2001)
2. Topolinski, T., Cichanski, A., Mazurkiewicz, A., Nowicki, K.: The relationship between trabecular bone structure modeling methods and the elastic modulus as calculated by FEM. Sci.World J. **2012**, 1–9 (2012). https://doi.org/10.1100/2012/827196. Article ID 827196

3. Cichański, A., Nowicki, K.: Trabecular bone microstructural FEM analysis for out-of plane resolution change. In: Arkusz, K., Będziński, R., Klekiel, T., Piszczatowski, S. (eds.) BIOMECHANICS 2018. AISC, vol. 831, pp. 210–218. Springer, Cham (2019). https://doi.org/10.1007/978-3-319-97286-2_19
4. Della Corte, A., Giorgio, I., Scerrato, D.: A review of recent developments in mathematical modeling of bone remodeling. Proc. Inst. Mech. Eng. Part H **234**, 273–281 (2020)
5. Sandino, C., McErlain, D.D., Schipilow, J., Boyd, S.K.: Mechanical stimuli of trabecular bone in osteoporosis: a numerical simulation by finite element analysis of microarchitecture. J. Mech. Behav. Biomed. Mater. **66**, 19–27 (2017)
6. Christen, P., et al.: Bone remodelling in humans is load-driven but not lazy. Nat Commun **5**, 4855 (2014)
7. Mulvihill, B.M., Prendergast, P.J.: Mechanobiological regulation of the remodelling cycle in trabecular bone and possible biomechanical pathways for osteoporosis. Clin. Biomech. **25**, 491–498 (2010)
8. Moreno, R., Borga, M., Smedby, Ö.: Techniques for computing fabric tensors: a review. In: Westin, C.-F., Vilanova, A., Burgeth, B. (eds.) Visualization and Processing of Tensors and Higher Order Descriptors for Multi-Valued Data. MV, pp. 271–292. Springer, Heidelberg (2014). https://doi.org/10.1007/978-3-642-54301-2_12
9. Bernard, S., Grimal, Q., Laugier, P.: Accurate measurement of cortical bone elasticity tensor with resonant ultrasound spectroscopy. J. Mech. Behav. Biomed. Mater. **18**, 12–19 (2013)
10. Cluzel, C., Allena, R.: A general method for the determination of the local orthotropic directions of heterogeneous materials: application to bone structures using mCT images. Math. Mech. Complex Syst. **6**, 353–367 (2018)
11. Sandino, C., McErlain, D.D., Schipilow, J., Boyd, S.K.: The poro-viscoelastic properties of trabecular bone: a micro computed tomography-based finite element study. J. Mech. Behav. Biomed. Mater. **44**, 1–9 (2015)
12. Cichański, A., Nowicki, K., Mazurkiewicz, A., Topolinski, T.: Investigation of statistical relationships between quantities describing bone architecture, its fractal dimensions and mechanical properties. Acta Bioeng. Biomech. **12**, 69–77 (2010)
13. Kim, D.G., Christopherson, G.T., Dong, X.N., Fyhrie, D.P., Yeni, Y.N.: The effect of micro-computed tomography scanning and reconstruction voxel size on the accuracy of stereological measurements in human cancellous bone. Bone **35**, 1375–1382 (2004)
14. Niebur, G.L., Yuen, J.C., Hsia, A.C., Keaveny, T.M.: Convergence behavior of high-resolution finite element models of trabecular bone. J. Biomech. Eng.-Trans. ASME **121**, 629–635 (1999)
15. Kothari, M., Keaveny, T.M., Lin, J.C., Newitt, D.C., Genant, H.K., Majumdar, S.: Impact of spatial resolution on the prediction of trabecular architecture parameters. Bone **22**, 437–443 (1998)
16. Nikodem, A.: Correlations between structural and mechanical properties of human trabecular femur bone. Acta Bioeng. Biomech. **14**, 37–46 (2012)
17. Luo, C., Liao, J., Zhu, Z., Wang, X., Lin, X., Huang, W.: Analysis of mechanical properties and mechanical anisotropy in canine bone tissues of various ages. Biomed. Res. Int. **2019** (2019). Article id: 3503152
18. Mazurkiewicz, A.: The effect of trabecular bone storage method on its elastic properties. Acta Bioeng. Biomech. **20**, 7 (2018)
19. Bevill, G., Keaveny, T.M.: Trabecular bone strength predictions using finite element analysis of micro-scale images at limited spatial resolution. Bone **44**, 579–584 (2009)

Optimization of Hip Implant Designs Based on Its Mechanical Behaviour

Hasan Göktaş[1], Eda Subaşi[1], Metin Uzkut[1], Mustafa Kara[1], Hamit Biçici[1], Hadi Shirazi[1], K. N. Chethan[2], and Şenay Mihçin[1](✉)

[1] Department of Mechanical Engineering, Izmir Institute of Technology, Urla, Izmir, Turkey
senaymihcin@iyte.edu.tr

[2] Department of Aeronautical and Automobile Engineering, Manipal Institute of Technology, Manipal Academy of Higher Education, Manipal 576104, Karnataka, India

Abstract. Total Hip Arthroplasty (THA) is one of the best advancements in healthcare. THA is required when the hip joint causes immobility and pain. The designed hip implants vary in geometry with different geometrical parameters. The geometry plays an important role in the mechanical behavior of the hip implant. In this study, the optimum selection of hip implant under static loading was evaluated using Finite Element Modeling (FEM). Hip implants with three different stem cross-sections including. (a) elliptic, (b) oval, and (c) trapezoidal were designed using a commercial Computer-Aided Design (CAD) software package. The FEM analysis was carried out via ANSYS R2019 to assess the key mechanical parameters of the implants such as stress distribution and deformation. The results were evaluated for the best stress and strain values. The optimum design had equivalent stress (von Misses) of 258,1 MPa, equivalent strain of 0.004, with total deformation of 0.24 mm and frictional stress of 0.362 MPa producing best values for trapezoidal cross-sectioned design. The findings of this study provided an insight into the selection of appropriate hip implant design with certain geometric design parameters to produce optimum results in clinical applications.

Keywords: Hip implant · Finite element method · Static analysis · von Mises stress · Total deformation

1 Introduction

Total Hip Replacement (THR) surgery is frequently applied when hip joints lose their functionality. To succeed in replacing a new hip prosthesis, alongside the appropriate surgery techniques, the implant must satisfy the requirements to fit for the purpose by exhibiting expected biomechanical behaviors such as optimum strength, wear resistance, osteointegration, and biocompatibility, etc. [1]. Stress distribution and deformation are the primary factors considered in the implant design research to shed light on the uniform (in correspondence with stress shielding effect) or detrimental unexpected stress-deformation, which are of great importance in the preliminary study on the mechanical performance of the implants before clinical application [2]. The common geometrical variables of the hip implants could be named as follows; stem length, cross-section type,

A. Hadamus et al. (Eds.): BIOMECHANICS 2021, LNNS 328, pp. 37–43, 2022.
https://doi.org/10.1007/978-3-030-86297-8_4

neck length, neck angle and head diameter. Titanium body metal hip implants are generally used without cement. The reason for this is that the titanium alloy implant makes too much displacement in the bone, which causes the femoral stem part to bend in the cement and destroy it [3]. Since Ti-6Al-4V alloy was used in the femoral stem part in 3 different cross-section designs in this study, cement was not used in fixing the hip implant into the femur bone [1]. The implants are designed with the expected lifetime of 10 years which can change according to the patient and their lifestyle [4]. The parameters such as equivalent stress, deformation, and friction play a role in this expected lifetime, depending on the used materials such as UHMWPE, Ti6Al4V, CoCrMo, and stainless steel [5, 6]. Stem design is an important aspect in hip implant assembly. Stem design requires optimization and needs to be investigated. The purpose of this study was to determine the effect of stem profiles in hip implant assembly by investigating their mechanical behaviour under certain loading conditions.

2 Materials and Methodology

Hip implants consist of 4 main parts which are femoral stem, femoral head, acetabular liner, acetabular cup [7]. Three different designs (shown in Fig. 1), which are representatives of commonly used hip implant models, were evaluated under static loading. Computer-Aided Design (CAD) drawings were completed in Solid Works 2016 (Dassault Systems, Velizy-Villacoublay, France) and were analyzed via ANSYS R2019 (ANSS, Canonsburg, Pennsylvania, USA) for Finite Element Modelling (FEM).

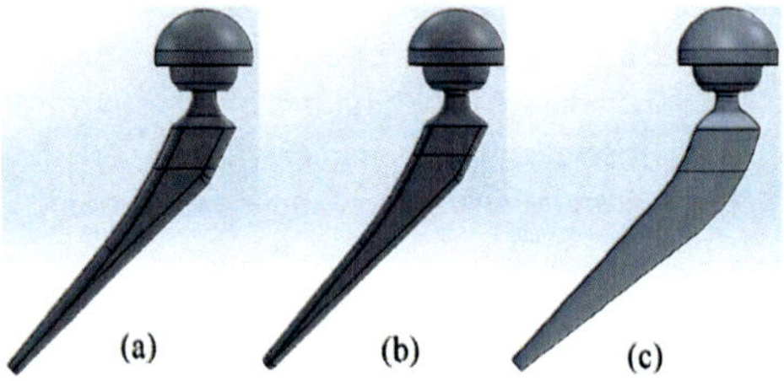

Fig. 1. CAD models of Hip Implants elliptic (a), oval (b), and trapezoidal (c).

Three main materials were chosen due to their mechanical behaviors, as shown in Table 1. In the designs, cobalt-chromium alloy was used for the femoral head and acetabular cup, Ti-6Al-4V was used for the femoral stem, UHMWPE was used for acetabular liner. The same material combinations were used within the 3 models. The cross-section is elliptic for design (a) and oval for design (b) and trapezoidal for design (c). Stem lengths, head diameters, neck lengths, neck angle kept constant for all of them. The stem length is 150 mm, the neck length is 32 mm and the neck angle is 135° [8]. The designs were subjected to static loading of 2.3 kN which is ISO 7206-4:2010 condition in the direction of gravity applied directly to the acetabular cup as shown in Fig. 2 [9]. The stem was selected as fixed support and the coefficient of friction of 0.15 was used between the acetabular liner and acetabular head [10, 11]. All degrees of freedom of stem area were constrained and the contacts between neck/head and liner/shell assumed to be bonded. The static structural analysis was evaluated according to the same boundary conditions using ANSYS R2019 and factors of safety were evaluated using ANYSY R2019 fatigue tool. In this study, the mesh size was used as 2 mm. The number of elements and nodes are listed in Table 2.

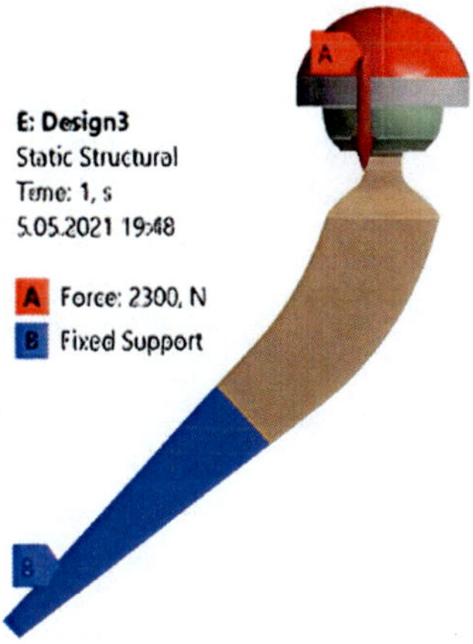

Fig. 2. Application point and application force.

Table 1. Mechanical properties of selected material [12].

Materials/properties	UHMWPE	Ti6Al4V	CoCrMo
Density (kg/m^3)	930	4430	8300
Young's modulus (MPa)	6900	115000	230000
Poisson's ratio	0.29	0.342	0.3
Yield strength (MPa)	21	880	612
Ultimate strength (MPa)	48	950	970

Table 2. Mesh size, element and nodes number of designs.

	Mesh size	Element	Node
Design a	2 mm	66204	103251
Design b	2 mm	67870	105510
Design c	2 mm	61284	108528

3 Results

As shown in Fig. 3, the contours of equivalent stress and strain, deformation, frictional stress and factor of safety were obtained for all three different designs. Additionally, the maximum values of these properties are listed in Table 3. As can be found from Table 3, the trapezoidal stem design could substantially decrease the max. stress and deformation values compared to the other designs. In addition, design (b) i.e. oval profile exhibited the max. stress and deformation values smaller than the design (a) i.e. elliptical one but bigger than design (c). The results also revealed that the changes in the stem profile had no significant effect on the factor of safety and frictional stress (<2%).

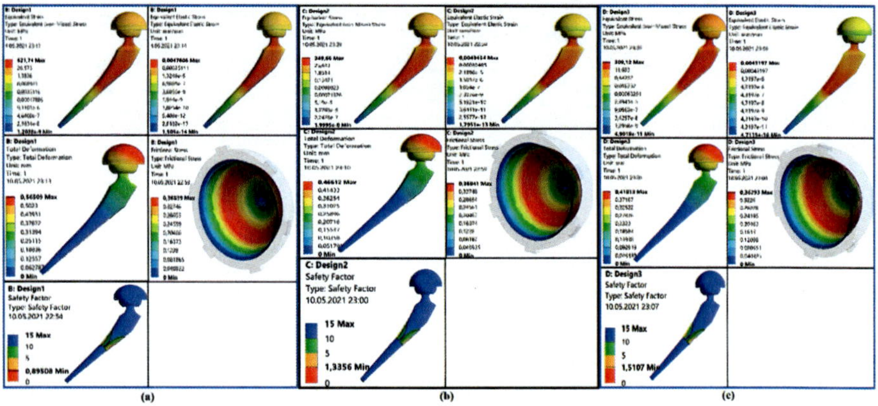

Fig. 3. Contours of the equivalent stress, equivalent strain, total deformation, frictional stress and safety factor of designs a, b, c.

Table 3. Mechanical results from ANSYS 2019.

		Equivalent stress (MPa)	Equivalent strain (mm/mm)	Total deformation (mm)	Frictional stress (MPa)	Factor of safety
Design a	Max	521,74	0,005	0,565	0,368	15
Design b	Max	349,66	0,004	0,466	0,368	15
Design c	Max	258,08	0,004	0,240	0,362	15

4 Discussion

In this study, elliptic, oval, and trapezoidal stem types were used as different geometry parameters. These three designs had the same dimensions but they differed in the stem shapes, shown in Fig. 1. In the previous studies, chromium-cobalt, titanium alloys, and UHMWPE have been reported as mostly used materials in hip implant systems because of their mechanical properties [13, 14]. Variations in stems' geometry were investigated to improve the design related to stem length, cross-section, and modified stems [15, 16]. In the literature, 285.1 MPa von-Misses stress was obtained when a force of 2.3 kN was applied to 175 mm stem length [17]. Another study reported 210 MPa von-Misses stress under ASTM F745 standards with 165 mm stem length [18].

The longevity of implants mainly depends on the structural strength. In this study, three different stem profiles were employed to recognize the optimum design. The models were successfully analyzed using ANSYS 2019 and the results were tabulated in figure & Table 3. Equivalent stress, total deformation and frictional stress were 258.08 MPa, 0.240 mm, and 0.362 MPa respectively for a trapezoidal design where these values were lesser than the results of the other designs. Our results showed the lower values of the max. stress, deformation and stress in comparison with the previous studies under the same frictional conditions [17].

In general, based on Table 3, max equivalent stresses were 521,74 MPa, 349,66 MPa, and 258,08 MPa for all designs. Max stress regions were mostly in the acetabular cup and femoral stem. According to their mechanical properties, Ti6Al4V has 880 MPa and CoCrMo has 612 MPa ultimate strength (see Table 1) [12]. The results were lower than these values.

Osseointegration is an important phenomena related to how bone is influenced after implantation. In this study, Ti-6Al-4V alloy is used and there is a research on different materials which are studied to understand how they are acting after implantation [19]. However, our study is a computational one limited with mechanical behaviour of implants, not including the tissue and bone integration aspects. Our study is designed for investigation of geometry aspects. Based on the obtained results, next step is to design experimental applications to investigate the affects of different geometries of stem profiles for osseintegration. The result of this computational study via ANSYS is a computational one, not focusing on tissue interface problems between the implant and the bones. In our next study with experimental focus, osseointegration aspect will be investigated for different geometry profiles.

5 Conclusions

In this study elliptic, oval and trapezoidal stem designs were modeled using ANSYS R2019 to investigate different stem cross-sectioned designs, consisting of Ti6Al4V stem and acetabular cup, cobalt-chromium head and PE liner. Based on equivalent stress and total deformation values, the trapezoidal design was chosen as the optimum model amongst all three designs under static loading conditions.

Acknowledgments. The authors thank the Mechanical Engineering, Department of Izmir Institute of Technology for providing the high computational facility to carry out this research.

Declarations.

Ethics Approval and Consent to Participate
Not applicable.

Consent for Publication
All authors consent to the publication of this manuscript.

Competing Interests
The authors declare that they have no competing interests.

Funding. This research was funded by TUBITAK 2232 International Outstanding Researchers Funding Scheme with Grant No of 118C188 'New Generation Implants for All' project.

References

1. Merola, M., Affatato, S.: Materials for hip prostheses: a review of wear and loading considerations. Materials **12**(3), 495 (2019)
2. Ridzwan, M., Shuib, S., Hassan, A., Shokri, A., Mohamad Ibrahim, M.: Problem of stress shielding and improvement to the hip implant designs: a review. J. Med. Sci. **7**(3), 460–467 (2007)
3. Mattingly, D.A.: Cemented long-stem femoral components in revision total hip arthroplasty. In: Bono, J.V., McCarthy, J.C., Thornhill, T.S., Bierbaum, B.E., Turner, R.H. (eds.) Revision Total Hip Arthroplasty, pp. 239–243. Springer, New York (1999). https://doi.org/10.1007/978-1-4612-1406-9_32
4. Ulrich, S.D., et al.: Total hip arthroplasties: what are the reasons for revision? Int. Orthop. **32**(5), 597–604 (2008). https://doi.org/10.1007/s00264-007-0364-3
5. Colic, K., Sedmak, A., Grbovic, A., Sedmak, S., Dordevic, B.: Finite element modeling of hip implant static loading. Procedia Eng. **149**, 257–262 (2016)
6. Wang, L., Isaac, G., Wilcox, R., Jones, A., Thompson, J.: Finite element analysis of polyethylene wear in total hip replacement: a literature review. Proc. Inst. Mech. Eng. Part H **233**(11), 1067–1088 (2019)
7. Baura, G.D.: Total hip prostheses. In: Medical Device Technologies, pp. 381–404 (2012)
8. Supra Art: hip prtohesis catolog. https://www.tstsan.com/urun/i/87/supra-art-1-stem.html
9. Kaya, F., İnce, G., Avcar, M., Yünlü, L. : Kalça protezi tasariminin sonlu elemanlar yöntemi ile statik analizi. Mühendislik Bilimleri Ve Tasarım Dergisi **9**(1), 199–208 (2021)
10. Damm, P., et al.: Friction in total hip joint prosthesis measured in vivo during walking. PLoS One **8**(11), e78373 (2013)
11. Ramoutar, D.N., Crosnier, E.A., Shivji, F., Miles, A.W., Gill, H.S.: Assessment of head displacement and disassembly force with increasing assembly load at the head/trunnion junction of a total hip arthroplasty prosthesis. J. Arthroplasty **32**(5), 1675–1678 (2017)
12. Rawal, B., Yadav, A., Pare, V.: Life estimation of knee joint prosthesis by combined effect of fatigue and wear. Procedia Technol. **23**, 60–67 (2016)
13. Darwich, A., Nazha, H., Daoud, M.: Effect of coating materials on the fatigue behavior of hip implants: a three-dimensional finite element analysis. J. Appl. Comput. Mech. **6**(2), 284–295 (2020)
14. Luo, C., et al.: Femoral stress changes after total hip arthroplasty with the ribbed prosthesis: a finite element analysis. BioMed Res. Int. 8 (2020)

15. Abdullah, A.H., Asri, M., Alias, M.S., Giha, T.: Finite element analysis of cemented hip arthroplasty: influence of stem tapers. In: Proceedings of the International Multi Conference of Engineers and Computer Scientists 2010, IMECS 2010, vol. III, 17–19 March 2010 (2010)
16. Tanner, K.E., Yettram, A.L., Loeffler, M., Goodier, W.D., Freeman, M.A., Bonfield, W.: Is stem length important in uncemented endoprostheses? Med. Eng. Phys. **17**(4), 291–296 (1995)
17. Griza, S., Reis, M., Reboh, Y., Reguly, A., Strohaecker, T.R.: Failure analysis of uncemented total hip stem due to microstructure and neck stress riser. Eng. Fail. Anal. **15**(7), 981–988 (2008)
18. Griza, S., Zanon, G., Silva, E.P., Bertoni, F., Reguly, A., Strohaecker, T.R.: Design aspects involved in a cemented THA stem failure case. Eng. Fail. Anal. **16**(1), 512–520 (2009)
19. Albrektsson, T., Johansson, C.: Osteoinduction, osteoconduction and osseointegration. Eur. Spine J. **10**(Suppl 2), S96–S101 (2001). https://doi.org/10.1007/s005860100282

Numerical Analysis of Scoliosis Brace

Slawomir Grycuk(✉) and Piotr Mrozek

Institute of Biomedical Engineering, Faculty of Mechanical Engineering, Bialystok University of Technology, Wiejska 45C, 15-351 Białystok, Poland
s.grycuk@doktoranci.pb.edu.pl

Abstract. In this article, results of numerical finite element method (FEM) simulations of a Boston orthopedic brace in the Ansys environment are presented. A reverse engineering methodology based on digitization by means of a three-dimensional (3D) optical scanner was employed to develop the geometric model. The force flow lines characterizing the brace and indicating the general working method of the orthosis's structure were determined using the FEM model. Identification of the main areas of the orthosis, carrying loads correcting the spine and of the positions of sites exerting little effort, from the perspective of their participation in the orthosis's essential therapeutic application, was carried out. Methods for mechanical optimization of the brace's design can be proposed based on the results obtained. As the conducted analysis is universal in character, it can be adapted to other types of orthopedic braces.

Keywords: Orthotics · FEM · Principal stress vector · Principal stress trajectories · Force flow lines

1 Introduction

Scoliosis is a severe, three-dimensional (3D) deformation of the spine [1]. Non-surgical treatment involving bracing, using the Boston brace [2], is the most commonly applied method of treatment. A brace can be described as a customized thoracic orthosis. Typically, it is relatively heavy and rigid. For treatment to achieve results, the brace must be worn for over 23h every day, practically from diagnosis to physical maturity of the body [3]. The apex of the spinal curve and support sites at two points on the opposite side are the sites of load application in the case of the simplest corrective action. It is called the three-point pressure system [4]. Multiple three-point corrective systems, applied onto a 3D space, are present in more advanced designs, e.g. the Cheneau brace [5].

Experiments associated with the determination of forces arising in components of the brace during corrective work, e.g. in brace tightening straps, or pressures acting on the brace, play a significant role in research concerning the structure of braces. For example, large numbers of pressure sensors can be placed as arrays on the interior side of the brace [6, 7]. Two directions of research can be distinguished. The first of them involves tests conducted with the involvement of the patient [6]. In the second direction, stationary stands for testing of forces with three-point brace loading are employed [8].

A. Hadamus et al. (Eds.): BIOMECHANICS 2021, LNNS 328, pp. 44–54, 2022.
https://doi.org/10.1007/978-3-030-86297-8_5

One can see that the progress made until now in the field of rigid braces concerns a better understanding of their corrective functions [4], described by, among others, the spatial field of forces applied to the torso required for correction, and, to a lesser extent, by the mechanical properties of braces needed for the performance of these functions. There are examples of FEM applications in the literature that describe individual braces as an additional option accompanying FEM modelling of the torso [9], but these orthosis models are simplified and do not represent the geometry of the actual orthosis sufficiently well.

Computer modelling of the brace's mechanical structure seems to be a difficult task due to its nature. It is thin-walled, has an open cross-section, and a spatial distribution of loads is applied to it. Attention has been paid in the literature to the relationship between the mechanical properties of the brace's material and the pressures exerted by the orthosis [1].More flexible braces are more comfortable to wear, but at the same time, more rigid orthoses are better at exerting corrective forces. There is literature concerning determination of corrective force distributions [3], but there is little information about the mechanical properties of the brace's structure, required for effecting these forces [7, 10]. Literature analysis shows that the values and directions of an orthopedic brace's corrective forces are critical factors in the efficacy of treatment by its means. Therefore, it is important to identify the conditions that the brace's mechanical structure must fulfill in order to ensure its capacity to implement the required field of forces.

One may hypothesize that the proper starting point for analysis of the mechanical features of existing brace designs is to develop a credible numerical model of an actual orthosis whose efficacy has already been confirmed in practice. In-depth modelling results of such structures are absent in the literature. The numerical simulations presented in this article were aimed at developing a reliable FEM model of a sample brace. In-depth analysis of the model will serve for definition of the working scheme of the brace's structure, which will then enable indication of the possibilities of optimizing the brace's design.

2 Materials and Methods

2.1 Research Object - Boston Brace

The Boston brace system enables non-surgical treatment of scoliosis by preventing scoliosis from progressing in growing patients [11]. Usually, the system is effective when treating curves with an apex between T-6 and L-3. Curves with apexes outside these limits generally cannot be treated effectively using a Boston brace [8]. Boston braces are applied to lumbar and sacrolumbar scoliosis, i.e. with apexes in lower regions of the spine. Due to this, the upper part of a Boston brace cannot be rigid and can only rest on the patient's armpit to prevent shifting of the orthosis.

A Boston brace for left lumbar curve treatment was selected for study. The brace was thermoformed from a 4-mm-thick polypropylene sheet. The orthosis's interior surface is lined with polyurethane foam. It is characterized by a three-point system of corrective forces, unlike other types of braces, e.g. the Cheneau brace, which is characterized by a spatial system of multiple corrective forces [12]. Applying a relatively simple system of forces enables simplification of the model and makes the numerical model more plausible.

2.2 Brace Geometric Model

A 3D geometric computer model of the brace was reverse engineered (3D scanning) based on the existing Boston brace presented in Fig. 1a. The orthosis underwent procedures to properly prepare it for scanning. As preparation for 3D measurement, the orthosis was stiffened by means of adjustable buckles at the level of the tightening straps (1). The brace's surface was coated with matting material (2), and reference points (3) enabling spatial correlation of subsequent 3D scans were glued on. The point cloud was created using an Atos Core 200 optical scanner manufactured by GOM GmbH (accuracy of measurement according to standard VDI 2634 Part 3 – no worse than 0.03 mm, positioning repeatability ± 0.05 mm, optical system with two CCD 5 Mpx cameras). After preliminary, rough processing, the point cloud yielded a mesh of STL triangles defined as a binary representation consisting of 2,953,425 elements. 33 exposures, using 25 markers, were applied for this purpose. Point cloud processing computer software, Geomagic Wrap (3D Systems), was used to remove the fastening buckles from the model, repair artifacts, and fill in losses. The brace's exterior surface was represented on the mesh modified in this way. Using this surface as the foundation, the next step was to create a shell with a thickness of 4 mm, corresponding to the structure of the brace, described by NURBS splines. The model representing the actual brace was created by processing the point cloud in this manner (see Fig. 1b). An interior pad, along with a corrective ring made of soft material, was also modelled at the height of the spinal curve's apex. The brace's computer model was imported into the ANSYS environment (ANSYS Inc.) for FEM numerical simulation, in which numerical analyses were then performed.

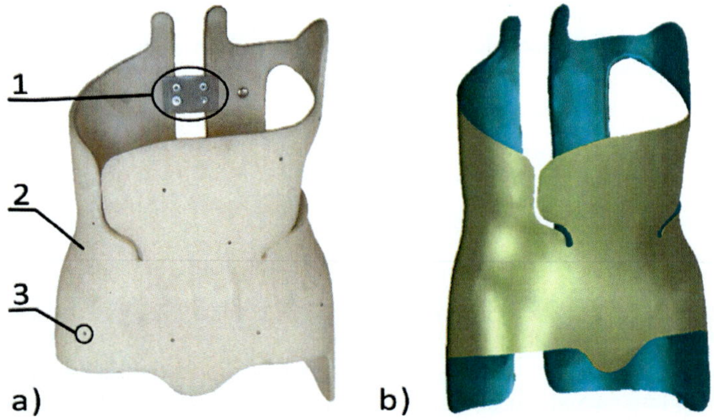

Fig. 1. Reverse engineering - creating a computer model of the orthosis based on a real object. Brace prepared for scanning (a): 1 – fastening stiffening the brace, 2 – matted exterior layer, 3 – reference markers. 3D model of the brace (b).

2.3 FEM Model

The material selected for the exterior body was polypropylene (E = 1000 MPa, v = 0.2) [6, 7], and soft foam was used to model the middle layer (E = 100 MPa, v = 0.45) [7]. The FEM model was generated using a 10-node tetrahedral element (SOLID187). A relatively fine mesh of tetrahedral finite elements (see Fig. 2a) was generated in order to show in detail the distribution of directions and values of the principal stress vectors, shown after solving the problem in each element separately. The shapes of the tetrahedrons appear quite regular, despite the complex geometry of the brace.

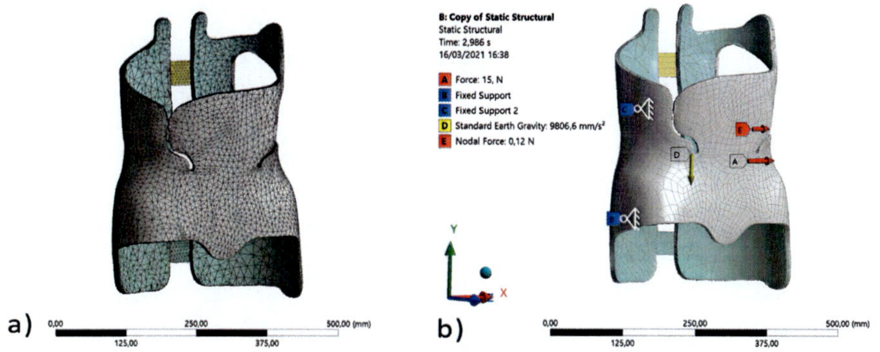

Fig. 2. Numerical brace model: generated mesh of tetrahedral finite elements (a), and boundary conditions on the imported geometric model represented by NURBS (b): preliminary force (A), fixed supports (B and C), additional force applied after preliminary loading of the brace (E), force of gravity (D).

The concentration of the mesh was also greater at critical sites with more complex geometry, so as to improve the accuracy of stress calculation when the brace was loaded at these sites. Discretization was performed until further concentration of the node mesh did not change the results of the von Mises stress obtained by more than 3%. After the mesh was concentrated, 149,322 nodes and 77,530 finite elements with a maximum edge length of 15 mm were obtained (see Fig. 2a).

After the finite element mesh was generated and the brace model's parameters were defined, the boundary conditions were introduced. Supports *B* and *C* and preliminary load *A* were applied to implement the three-point system of forces (see Fig. 2b). Small surfaces *B* and *C* were defined as two immobile supports, fixed in space. Force *A*, with a value of 15 N, was applied as the preliminary load of the brace. The value of force A falls within the range of real-life forces occurring in braces [13]. Force of gravity *D* also acts on the brace when it is immobilized in this manner. In order to best reflect the actual conditions of the experiment, the force of gravity was taken into account, although it does not seem to play a big role. The field of displacements was calculated for the loads and supports defined in this way. Following the first stage of calculations, described above, an additional force *E*, with a value of 0.12 N, was applied to the brace in the same direction as applied force *A*. The purpose of introducing the force *E* was to obtain numerical results of displacements that could be directly compared with the results of experimental tests of the brace, carried out on an interferometric test stand, which, however, were not included in the scope of this article. As shown further on in the article, the brace's material works within the linear range for typical loads acting on the brace. Therefore, test results of small displacements at low loads have a linear relationship with results corresponding to greater, operational loads acting on the orthosis. The final result of simulations was the difference in displacements before and after the input of force *E*. The distribution of displacements on the *Z* axis was obtained as a result, which can be verified in the future with the results obtained on the experimental test stand.

3 Results

The brace's distribution of displacements, being the result of FEM numerical analysis, is shown in Fig. 3. In the numerical simulations, the maximum displacement on the Z axis was $-1.863\ \mu\text{m}$.

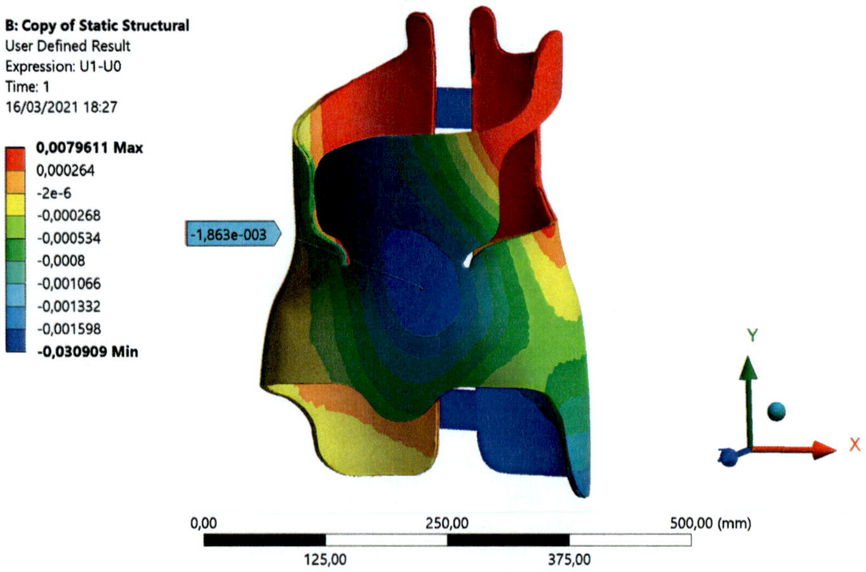

Fig. 3. Displacement distribution on the Z axis obtained by FEM numerical analysis.

Von Mises stress distribution as shown in Fig. 4 was obtained. With the exception of relatively small areas where loading force and supports were applied, the value of stresses in most of the brace was less than 2.878 MPa (see Fig. 4), falling within the linear range of the stress–strain curve for polypropylene [14], which confirms the previously adopted assumptions pertaining to the brace model.

Clearly, it is difficult to determine the brace's working scheme based on Fig. 4 alone. To illustrate how the brace's mechanical structure carries loads more clearly, Fig. 5 shows a view of the brace's front part, presented in a way that makes it possible to illustrate the vectors of the principal stresses on the interior and exterior of its front wall.

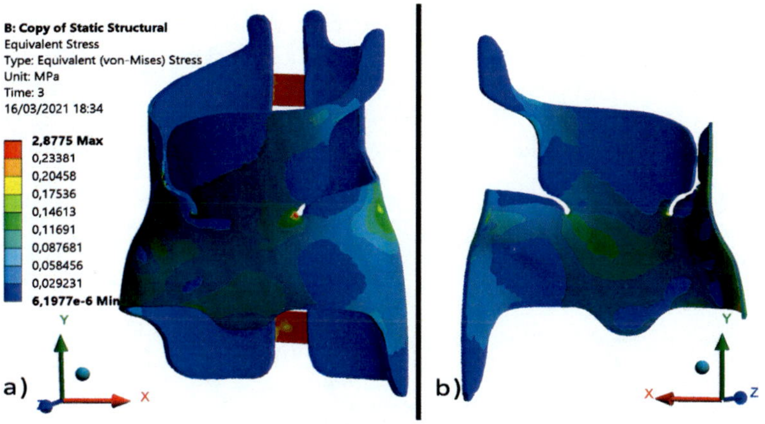

Fig. 4. View of brace with von Mises stress values: front exterior (a), and front interior (b).

Figure 6 shows the vectors of principal stresses for tensile stresses σ_1 only. The trajectories of maximum principal stresses σ_1 are also plotted in this figure. Lines starting and terminating in areas where external forces were applied were drawn along the directions in which stress vectors acted. The trajectories of these lines formed splines tangent to σ_1 vectors, with mild changes of direction.

4 Discussion

In this approach of simulating the brace's work, the standard method of analyzing FEM modeling results was applied, involving determination of the von Mises stress distribution for loads with values equal, in terms of orders of magnitude, to the typical corrective forces applied according to the elementary three-point pressure system. The results given in Fig. 4 indicate that the adopted thickness of the polypropylene shell is what gives the orthosis the required strength parameters. The allowable stress (F_α) values for this material are substantially greater than the von Mises stress values obtained. With the exception of the sites at which force and supports were applied, stress values in other parts of the brace were below 2.878 MPa, which corresponds to the linear range of the stress–strain curve for polypropylene [14]. Both sites of the brace most exposed to damage in mechanical terms and areas performing almost no work can be indicated. Nevertheless, it would be difficult to generalize the results obtained to braces under different load values and having different geometries based on the rather irregular distribution of stress values in the body of the brace (see Fig. 4), even if a similar three-point pressure system were to be applied.

Seemingly, the distributions of principal stresses, shown in different parts of the structure in Fig. 5, can provide more information about the general character of the orthosis's work. It can clearly be seen that the exterior surface of the brace's front wall is subject to compression, whereas the interior surface of the same wall is subject to tension. The absolute stress values in the exterior layer are about half those on the interior side. The work of other parts of the brace's shell is similar. In the predominant part of the

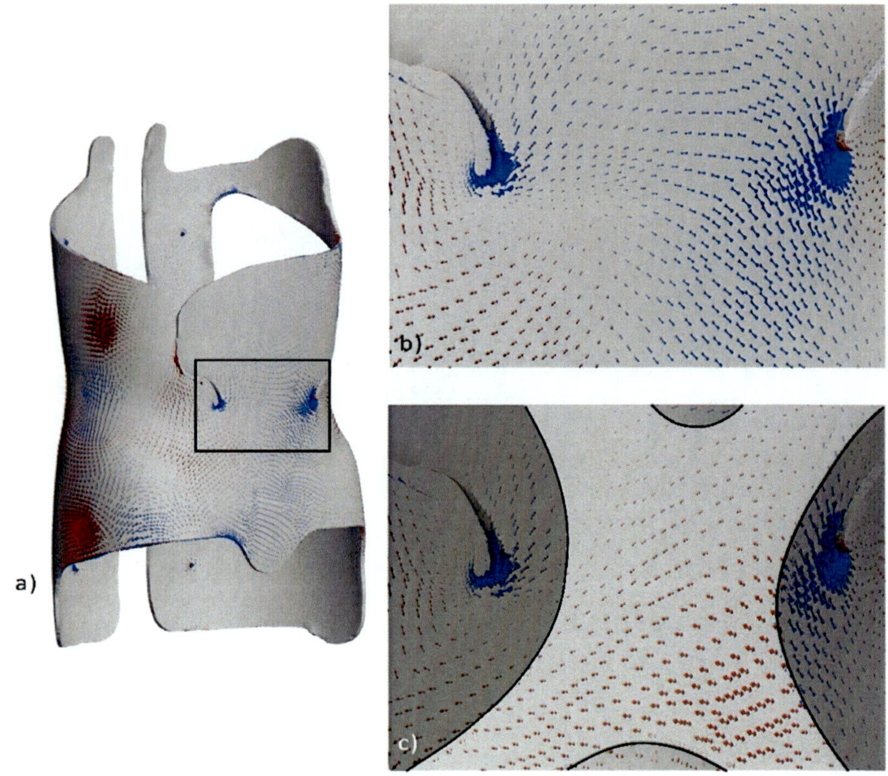

Fig. 5. Distribution of principal stresses σ_1 (red) and stresses σ_3 (blue) in the area of the brace's front wall marked in figure: area of the brace's front wall (a), stress distribution on exterior surface of marked area (b) and stress distribution in cross-section of front wall (c); the layer close to the interior surface is visible after "removal" of a part of the front wall's material.

brace shell, the characteristic stress distribution features a substantial value of tensile stresses on the interior side, arising from the superposition of stresses originating from two separately acting types of loads. By comparing the signs and values of the acting stresses, one may conclude that the brace's shell is subject to a complex state of bending and tension (see Fig. 7). This observation is of fundamental significance in the context of the method of brace optimization described further on in the article, since it enables determination of which side of the brace's wall will be subjected to the maximum tensile stresses.

Based on the determined vector fields of principal stresses and the plotted trajectories of maximum principal stresses σ_1 in areas where these stresses accept relatively high values (see Fig. 6), one can see clear regularities shaping the obtained image of the lines. The principal stress trajectories are arranged in a characteristic manner, connecting sites of corrective load application along paths, with relatively small lengths, in sum total. This picture corresponds to the concept of force flow lines, which is under intense scrutiny in optimal design theory [15]. This concept bases on the observation that, in the general case, the "flow" of forces through a mechanical structure between points of external load

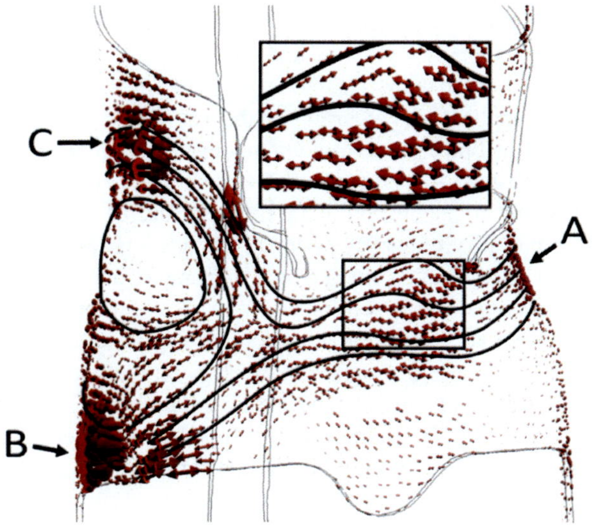

Fig. 6. Distribution of principal tensile stress vectors with principal stress trajectories.

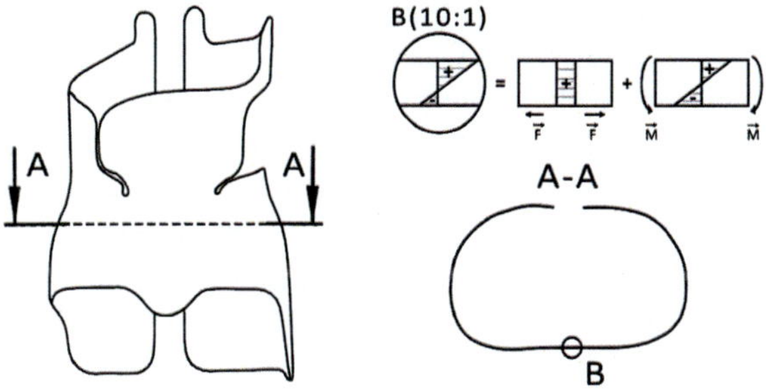

Fig. 7. Stress distribution diagram in cross-section A-A of the brace's front wall (the top side of diagram B corresponds to the interior side of the brace wall): superposition of stresses originating from tension under force F and bending by moment M.

application is most intensive in the vicinity of topographic lines connecting these points in space, considering the geometric boundaries of the object. Principal stress trajectories are one among many methods of implementing the force flow lines concept. There are, however, other, competing propositions in the literature [16]. Based on this concept, a brace can be described as a spatial mechanical system, mainly carrying loads spanning relatively short trajectories between sites of corrective load application. Based on the spatial distribution of the trajectories shown in Fig. 6, the area of the brace responsible for realizing the orthosis's corrective function can be indicated. It is located in the area of the plotted trajectories. Other areas serve an auxiliary function, e.g. stabilizing the brace with respect to the torso. The position of the corrective area is similar to the one determined using the optimization procedure [6]. Nonetheless, it seems that the method of analyzing the structure's work, as presented above, is more easily comprehensible in the context of stress distribution analysis compared to the approach based on minimization of elastic energy in the topological optimization method.

Based on the results of the numerical model presented above, one can indicate methods of their practical application. According to the test results, one can conclude that it is possible to reduce the brace's weight without substantially changing its mechanical parameters, particularly its stiffness in areas carrying corrective loads. Manufacturing the brace using a thinner material, with additional stiffening in area of the plotted trajectories shown in Fig. 6, responsible for carrying corrective loads, may be the simplest solution. Within this area, the brace can be stiffened by, for example, integrating thin overlays of composite, with a polymer matrix reinforced by glass or carbon fibers oriented along the trajectories of principal stresses shown in Fig. 6, into its surface. A comparison of Figs. 5, 6 and 7 clearly shows that applying overlays on the interior side of the brace in its front part would be most effective, because the interior side is under substantially greater load than the exterior side in this area. The opposite situation occurs near corrective force application sites. There, the exterior side carries much greater loads than the interior side. One can indicate sites for application of overlays in other areas carrying corrective loads in similar fashion.

The discussion above shows that the trajectories of force flow lines in the brace, serving as the basis for determination of areas requiring reinforcement, can be predicted rather easily, in the case of a three-point system of forces. It seems that such an approach can also be applied in more complex cases. In the simplest case, Boston brace uses a three-point pressure principle as the method of correction, which involves fixation above, below, and at the apex of the curve [17]. The chosen simulation setup is close to actual, physiological loading. For more complex cases, numerical calculations should take into account the appropriate, more complex physiological load systems. It may cause some limitations of the presented method of analysis, but it seems that the described procedure may be appropriate to optimize brace design also in these more complex cases.Verification of this hypothesis requires additional numerical simulations, however. Another way of practically applying the results of analysis is to remove material, e.g. by making holes in areas of little significance from the point of view of the brace's corrective properties.

The current trend in research on braces involves searching for a material that would serve as a substitute for polypropylene and enable FDM printing [18]. The results obtained fit perfectly into this trend, as they may serve as the foundation for a method of implementing a 3D-printed structure based on an openwork geometry corresponding to the distribution of principal stress trajectories and fulfilling the condition of minimal weight alongside the required stiffness of the structure [6].

The obtained results of this research may serve as the basis for further work leading to optimization of the brace's design through its stiffening and mass reduction. Considering the conclusions drawn from modeling results, particularly regarding the distribution of principal stress directions and values, stiffening components, working according to the determined directions of principal stresses, can be positioned appropriately. At the same time, the brace's mass can be reduced by reducing the amount of material in areas where the orthosis is under lesser load.

5 Conclusion

The brace's working scheme was defined by determination of the area of the orthosis subjected to mechanical loads, realizing its corrective function. Stress distributions in the brace's structure were determined, and trajectories of principal stresses were plotted on their basis, illustrating the "flow of forces" in the simple three-point pressure system. In this way, it was possible to identify sectors of the orthosis of particular significance from the perspective of corrective action. It seems that using the FEM model for further tests, with different variations of loads, and with values and distributions similar to those occurring in reality, may serve as a basis for drawing conclusions concerning the working method of the real-life orthosis. The correctly formulated computer model is the foundation for conducting numerical simulations with a high degree of reliability, with the ultimate objective of optimizing the brace's structure. The analysis under discussion is universal in nature and can easily be adapted to other types of braces.

References

1. Chan, W.-Y., et al.: Mechanical and clinical evaluation of a shape memory alloy and conventional struts in a flexible scoliotic brace. Ann. Biomed. Eng. **46**(8), 1194–1205 (2018). https://doi.org/10.1007/s10439-018-2016-8
2. Raux, S., Kohler, R., Garin, C., Cunin, V., Abelin-Genevois, K.: Tridimensional trunk surface acquisition for brace manufacturing in idiopathic scoliosis. Eur. Spine J. **23**(4), 419–423 (2014). https://doi.org/10.1007/s00586-014-3337-4
3. Chan, A., Lou, E., Hill, D., Faulkner, G.: Design and validation of transducers to measure interface force distribution in a spinal orthosis. Med. Eng. Phys. **34**, 1310–1316 (2012)
4. Rigo, M.D., Villagrasa, M., Gallo, D.: A specific scoliosis classification correlating with brace treatment: description and reliability. Scoliosis**12,** 1–11 (2010). https://doi.org/10.1186/1748-7161-5-1
5. Rigo, M., Negrini, S., Weiss, H., et al.: SOSORT consensus paper on brace action: TLSO biomechanics of correction (investigating the rationale for force vector selection). Scoliosis **5**, 16857045 (2006)
6. Liao, Y.-C., Feng, C.-K., Tsai, M.-W., et al.: Shape modification of the boston brace using a finite-element method with topology optimization. Spine **32**, 3014–3019 (2007)
7. Périé, D., Aubin, C.E., Lacroix, M., et al.: Biomechanical modelling of orthotic treatment of the scoliotic spine including a detailed representation of the brace-torso interface. Med. Biol. Eng. Comput. **42**, 339–344 (2004)
8. Chung, C.L., Kelly, D.M., Steele, J.R., DiAngelo, D.J.: A mechanical analog thoracolumbar spine model for the evaluation of scoliosis bracing technology. J. Rehabil. Assist. Technol. Eng. **5**, 205566831880966 (2018)
9. Guan, T., Zhang, Y.: Determination of three-dimensional corrective force in adolescent idiopathic scoliosis and biomechanical finite element analysis. Front. Bioeng. Biotechnol. **8**, 1–13 (2019)
10. Nijssen, J.P.A., Radaelli, G., Herder, J.L., et al.: Design and analysis of a shell mechanism based two-fold force controlled scoliosis brace. In: A: 41st Mechanisms and Robotics Conference. American Society of Mechanical Engineers, vol. 5, Cleveland, Ohio, USA, p. V05AT08A014 (2017)
11. Fortin, D., Cheriet, F., Beauséjour, M., et al.: A 3D visualization tool for the design and customization of spinal braces. Comput. Med. Imaging Graph. **31**, 614–624 (2007)

12. Rigo, M., Jelačić, M.: Brace technology thematic series: the 3D Rigo Chêneau-type brace. Scoliosis **12**, 10 (2017)
13. Gignac, D., Aubin, C., Dansereau, J., Labelle, H.: Optimization method for 3D bracing correction of scoliosis using a finite element model. Eur. Spine J. **9**, 185–190 (2000)
14. Khlif, M., Masmoudi, N., Bradai, C.: Polypropylene tensile test under dynamic loading. J. KONES. Power. Transp. **21**, 132-138 (2016)
15. Czarnecki, S., Wawruch, P.: Construction of stress trajectories in optimal, non-homogeneous elastic bodies. In: Advances in Mechanics: Theoretical, Computational and Interdisciplinary Issues, pp. 137–140 (2016)
16. Kelly, D.W., Tosh, M.W.: Interpreting load paths and stress trajectories in elasticity. Eng. Comput. **17**, 117–135 (2000)
17. Boston Brace International, Inc.: Reference Manual for the Boston Scoliosis Brace, Boston Brace International, Inc., Boston (2003)
18. Ng, K.J., Duke, K., Lou, E.: Investigation of future 3D printed brace design parameters: evaluation of mechanical properties and prototype outcomes. J. 3D Print. Med. **3**,171–184 (2019)

Small Unsecured Objects Transported in a Vehicle and Their Impact on Human Head Injury– Blunt Injury Criterion Approach

Jaroslav Hruby[1](✉), Brad Parker Wham[2], Zdenek Krobot[3], and Marek Semela[1]

[1] Institute of Forensic Engineering, Brno University of Technology, Brno, Czech Republic
jaroslav.hruby@colorado.edu

[2] Center for Infrastructure, Energy and Space Testing, CU Boulder, Boulder, CO, USA

[3] Department of Special and Combat Vehicles, University of Defense, Brno, Czech Republic

Abstract. The presented study deals with a familiar situation when unsecured objects are transported in a vehicle. Unsecured objects have a character of objects dedicated to daily usage – work purposes or personal purposes. That generally means laptops, cell phones, tablets, drinks in a glass bottle, objects for sports, and others. Objects of interest are considered stiff/rigid with an insignificant portion of deformability.

The study focuses on the interaction between unsecured objects placed inside a vehicle and vehicle occupants – if a traffic accident happens. If an unsecured object is randomly placed inside the vehicle's inner structure and the vehicle crashes into the barrier, the unsecured objects act like projectiles. These projectiles may, in some cases, interact with occupants' bodies. The interaction may cause, in specific cases, a severe occupant injury. Regarding the human body, the critical part taken for the study purposes is a human head – respectively rear part of a human head with a theoretically insignificant skin thickness. The blunt injury potential (when an unsecured object interacts with a human head) is calculated through Head Injury Criterion (HIC) and Blunt Criterion (BC). Blunt Criterion plays in the presented study a significant role because it serves as a HIC comparison and verification, and the inputs to Blunt Criterion must be carefully selected. If not, the correlation between Head Injury Criterion and Blunt Criterion is not adequately justified.The presented case study shows the selection of the proper and improper values for the Blunt Criterion computations and the influence of the selected values on obtained Blunt Criterion results.

Keywords: HIC · Blunt criterion · Projectile · Human head · Impact

1 Introduction

The primary commercial vehicle mission is to transport vehicle occupants between different locations securely and with suitable comfort. People (vehicle occupants) use vehicles for comfortable transportation between their homes and work, between their

A. Hadamus et al. (Eds.): BIOMECHANICS 2021, LNNS 328, pp. 55–70, 2022.
https://doi.org/10.1007/978-3-030-86297-8_6

homes and a vacation location or supermarket. Each transportation situation has a secondary purpose (primary is the occupant transportation) which includes transportation of typically unsecured objects of different usage.

If vehicle occupants transport themselves into the work, they also transport laptops, tablets, cell phones, or small drinks. If the transportation situation has a vacation purpose, vehicle occupants transport inside the vehicle tablets, notebooks, food, sports equipment, and indoor or outdoor apparel. Finally, for the shopping purpose (transportation from the supermarket), occupants transport inside the vehicle purchased goods (food, shoes, electronics) from the supermarket back to their homes.

The transported objects are commonly placed into the vehicle structure (inside the vehicle) randomly in random places (all around the vehicle's inner structure). Modern vehicles have many safe places in their inner structure, where these objects of everyday usage can be placed, but the objects mentioned above are typically not placed there. This situation is happening due to people's stress and tight schedules in their work or personal life. Safe places are commonly situated in the modern vehicles in the vehicle trunk. The transported objects are fastened in these places by plastic barriers or a safety system consisting of a safety net with hooks. Inner vehicle structures may also have some safe storage places, but it is not common to have a larger storage space in the vehicle occupant area. Objects are placed in many cases on the front co-driver seat, rear seats, or backplate situated above rear seats. If a traffic accident happens, the randomly placed objects inside a vehicle's inner structure may act as projectiles. The "projectile" behavior of objects is typical for frontal vehicle impacts with a barrier, where the vehicle deflection/impact area is enormous but with a small breaking distance - this means that the vehicle stops its motion on a short distance (let us say 0.5 to 1 m). If the vehicle stops at a very short distance, the deceleration acting on the vehicle and the objects inside is significant. A common situation that can describe the case mentioned above is vehicle impact with heavyweight truck, wall of the building, tree, or another vehicle in full range (100% coverage). Vehicle breaking actions may also play an important role when unsecured objects act as projectiles, but the braking distance is in many cases long enough - the objects do not play the same hazardous role in vehicle occupant safety as for the vehicular crash event. The study is focused primarily on vehicular frontal crashes with a barrier. Small unsecured objects are placed, for example, on the plate above rear vehicle occupant seats(or in the vehicle trunk) – see Fig. 1. The study deals with blunt human head injury potential - unsecured object interaction with the rear part of the human head, dismissing the head restraint system's presence and interaction of the human head with inner parts of the vehicular structure.

The unsecured objects are considered stiff/rigid with no significant sharp pats or edges – small laptops, tablets, cell phones, and other similar objects. The experimental approach is used to determine the head injury potential.

A combination of Head Injury Criterion (HIC) and Blunt Criterion (BC)determines human head injury potential. Focus is mainly taken on the Blunt Criterion usage with different input calculation values regarding mass and anthropometric data. The anthropometric data play a significant role in Blunt Criterion determination, and their variations may change the results significantly.

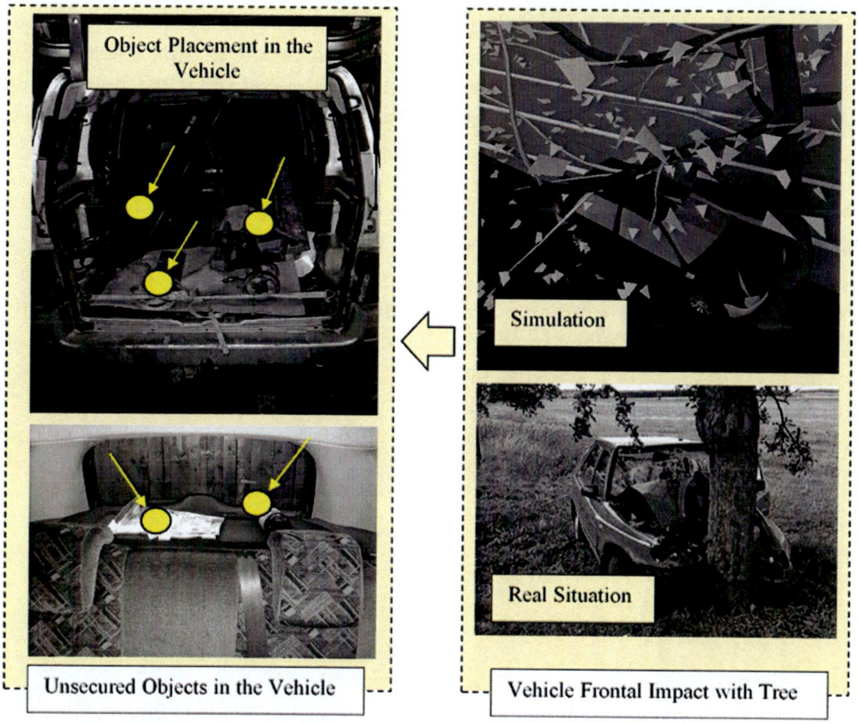

Fig. 1. Placement of the unsecured objects randomly in the inner structure of the vehicle and vehicular frontal crash event description.

2 Materials and Methodology

Experimental and analytical approaches were used for head injury potential determination through Blunt Criterion (BC). Head Injury Criterion (HIC)is used as a verification tool. The analytical injury potential calculations (HIC and BC) are based on the experimental data obtained through calibrated Hybrid III 50th Percentile Male Dummy under tests. Blunt Criterion values are also tied to the experimental data achieved from projectile movement, projectile shape, and anthropometrical data based on studies performed previously for different research purposes related to wound ballistics [1, 2].

2.1 Experimental Setup

The experimental part of the study was conducted in laboratory conditions using a self-developed airgun, including pressurizing system (pressure up to 10 bar) and a safety system. Self-developed projectiles were used with mountable mass for the impact initialization process. As a human model, Hybrid III 50th Percentile Male Dummy for frontal vehicle crash usage was chosen. The dummy was in a sitting position using a vehicle seat without a head restraint system and including a safety belt (body restraint system). The airgun for experimental investigation purposes mainly consists of a steel barrel, optical barriers, and pressurizing system. The steel barrel is made out of a could-welded steel tube with a polished inner surface to decrease the friction effect produced by the moving projectile. One side of the barrel tube has a dead end with a pressure input hose; the second side is open to let the projectile get out of the barrel. The basic concept of the airgun can be seen in Fig. 2.

Optical barriers are measuring the projectile velocity as the projectile leaves the barrel. Optical barriers are placed close to the open barrel side at a fixed distance of 100 mm (satisfactory projectile velocity determination). The optical principle of the barriers was selected because if we use a mechanical-based principle, a loss of functionality occurs after a specific amount of time (in-service) – deformation of spring elements holding the metal sheet triggers. When performing experimental verification of the airgun and measuring barriers performance (mechanical vs. optical), the mechanical barriers were damaged after approximately twenty shots.

The pressurizing system consists of two valves and a storage tank. Connections between individual pressurizing parts consist of hoses and tubes (with a limited pressure level up to 10 bar).

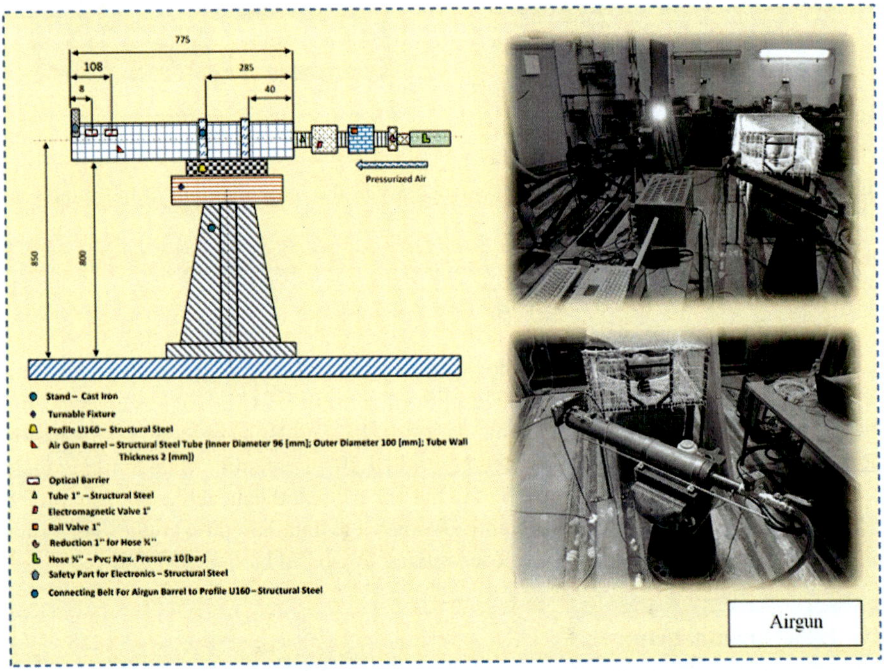

Fig. 2. The basic concept of the projectile used for the experiment (dimension values are in mm).

Projectiles are manufactured from nylon, steel, and hard rubber. The body of the projectiles is made out of nylon (friction elimination of the projectile in the barrel). Inner projectile part ensuring the weight diversity (changeable mass through mounting) is made out of steel (bolts, mountable mass). The impact projectile part is made out of hard rubber (to ensure that the Hybrid III 50th Percentile Male head (soft tissue part) will not get damaged by the projectile impact. The projectiles can have different weights from 0.5 kg up to 1.5 kg – depends on the weight of the steel part mounted in the nylon projectile body. Projectile stiffness characteristics are presented in Fig. 3 – compression test performance. The projectile concept can be seen in Fig. 4.

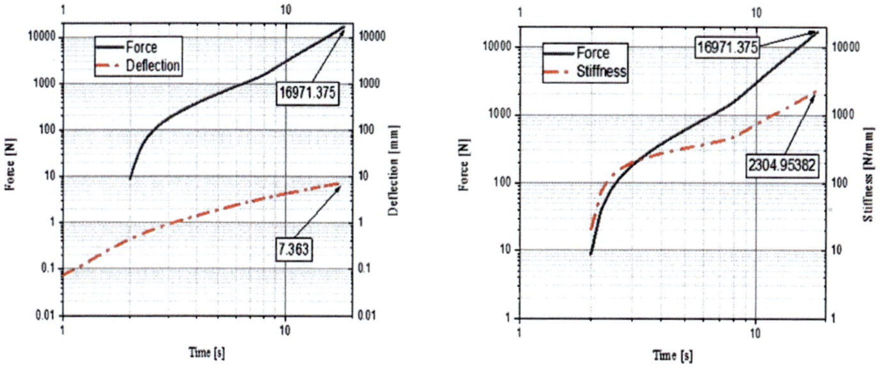

Fig. 3. Stiffness characteristics of the projectile based on the performed compression test.

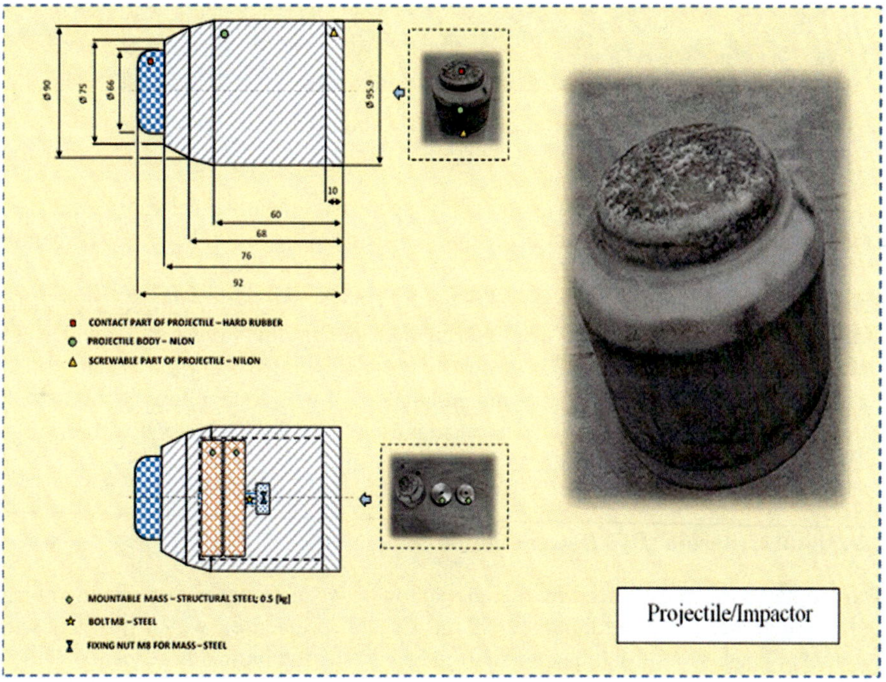

Fig. 4. The basic concept of the projectile used for the experiment (dimension values are in mm).

Hybrid III 50th Percentile Male Dummy is a critical part of the experimental setup. It is a certified and calibrated component of the experimental chain mainly used in the automotive industry to measure the severity of the frontal vehicle impact on a barrier (vehicle occupant safety). The Hybrid III 50th Percentile Male Dummy has built-in measurement apparatus. Data acquisition is performed through a separate device.

The Hybrid III 50th Percentile Male Dummy was placed/seated in the vehicle seat from Volkswagen Sharan without a head restraint system. The head restraint system was not present due to the conservative approach applied for achieving the maximum head injury potential and the proper shape of the impulse function. A body restraint system was applied in the specific position as it is in the vehicle. The body restraint system does not play a significant role in the experiment and is used for securing the Hybrid III 50th Percentile Male Dummy in the seat in a steady-state position. The positioning of the Hybrid III 50th Percentile Male Dummy can be seen in Fig. 5.

Fig. 5. Hybrid III 50th Percentile Male Dummy – Position during the test.

The data acquisition (dummy head response measurement - acceleration) was performed through the internal data acquisition system of the Hybrid III 50th Percentile Male Dummy with a sampling frequency equal to 100 kHz. Obtained acceleration data were loaded to a computer with an installed NI DIADEM Crash Analysis Toolkit for further analysis using CFC1000 data filtering and specific functions for the final acceleration impulse achievement. The electrical signals obtained from the airgun optical barriers were measured through instrument DEWE A4 with a sampling rate of 10 kHz. After each measurement of the electrical impulses from the airgun optical barriers, a calculation of projectile velocity was performed using DEWESOFT version 7.

2.2 Blunt Criterion (BC) Determination

The estimated injury produced by the interaction between an unsecured moving object and a human head can be defined as blunt. Object (projectile), which interacts with the human head, is stiff and has no sharp edges, so as the human head rear part. The interaction situation on which the blunt criterion calculation (also HIC) is based can be seen in Fig. 6.

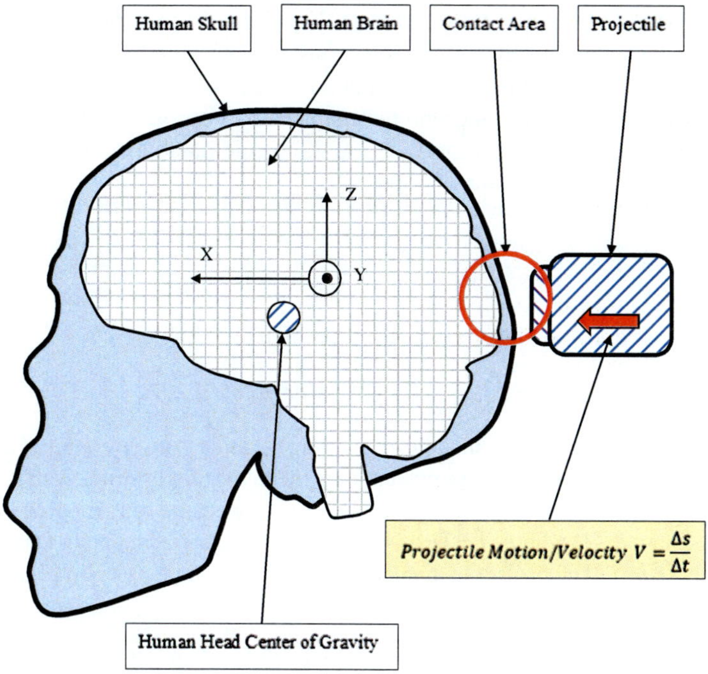

Fig. 6. Human head and projectile/object contact situation.

Blunt criterion can be calculated through Eq. (1) which can be seen below. The equation is adapted from the studies presented by Frank et al. [1] and Sturdivan [2]:

$$Blunt\ Criterion = ln\left(\frac{{}^{1}\!/_{2} \cdot M \cdot V^2}{W^{1/3} \cdot T \cdot D}\right) \tag{1}$$

The numerator in Eq. (1) represents the kinetic energy of the projectile. The kinetic energy is partially transformed to deformation energy and pulse/post-pulse movement of the human head and projectile. The symbol M is representing the total mass in kg of the projectile. Symbol V (squared) represents projectile velocity in m/s measured from the airgun optical barriers – barriers are placed at a constant distance of 100 mm, and we know the time difference from each projectile pass. The numerator input values are strongly dependent on the projectile characteristics regarding its motion. Values of the projectile mass were 0.5 kg, 1.0 kg, and 1.5 kg. Projectile velocity V values are the pressure functions applied to the airgun barrel through valves and airgun pressurizing system.

The denominator in (1) is representing variables that must be carefully selected regarding the impact situation. Symbol W defines the mass of the weight of the struck individual body part. Study [2] uses a mass for most of the whole body (impact on human chest). Study [1] is using (compared to [2]) differentiation of the struck body masses – head impact uses head mass, and other. The proper W definition is playing a significant role in blunt criterion calculation. Symbol T represents the combined thickness of the struck human body part at the specific location (impact area). The T value (in cm) depends on the anthropometric data and affects the resultant blunt criterion value. Parameter D represents the diameter of the projectile in cm. According to Sturdivan [2], the D value uses two forms. One form uses the modified diameter of the projectile; another form uses projectile diameter without modification. The projectile modification process is based on Eq. (2) adapted from the study [1] and [2]:

$$D_{(MODIFIED)} = 2 \cdot \left(\frac{A}{\pi}\right)^{1/2} \tag{2}$$

Parameter A in Eq. (2) can be calculated through Eq. (3):

$$A = \pi \cdot T \cdot (D - T) \quad where\, D > 2 \cdot T \tag{3}$$

The projectile diameter modification regarding (2) and (3) is only valid; if the projectile diameter D is twice or more significant, then the human body struck wall thickness T [2]. For the computations of blunt criterion in this study is used only modified projectile parameter (input to (1) where $D = D_{(MODIFIED)}$) – the projectile diameter D is more than twice larger than T. The projectile diameter modification can be seen in Fig. 7.

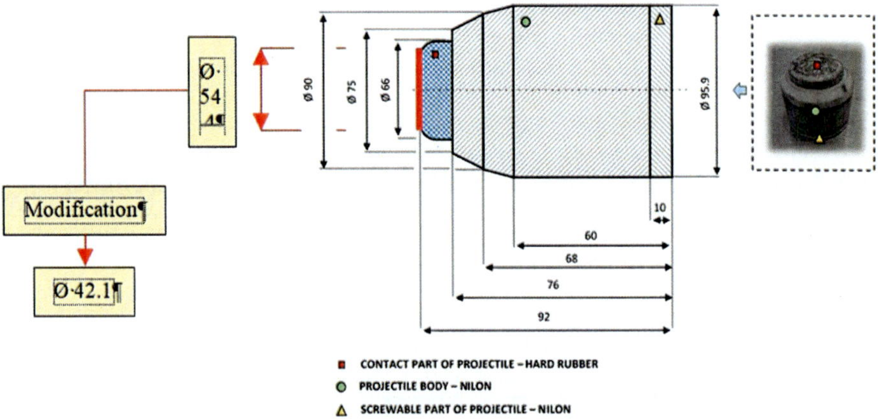

Fig. 7. The projectile diameter modification (dimension values are in mm).

The specific values of T for the blunt criterion calculation suggested for careful selection are in Table 1. These values are also used for the blunt criterion calculations presented in this study. The values of T were adapted from [1, 3]. For this case, when the projectile interacts with the rear part of a human head with a small portion of soft tissue, the relevant thickness value can be T = 1 cm or less. Other T values are dependent on human age [3]. Sincethe male dummy was used for the experimental injury investigation, T values adapted from [3] are related to the male gender and are random (left and right side of a human head), not specific for the human head rear part.

Values of W (see Table 2) were considered 4.54 kg, and 75.39 kg for Blunt Criterion comparative calculation purposes [5].

Table 1. Specific values for T parameter used for blunt criterion calculations

Calculation Case No	Values of T in cm	Adapted from
1	1	[1]
2	1.13	[6]
3	1.36	[6]

A value of Blunt Criterion equal to 1.6 was selected as critical regarding the study presented in [1]. The value 1.6 represents a 50% probability of skull fracture [1].

As a comparison tool to Blunt Criterion was chosen Head Injury Criterion (HIC). The critical value of the Head Injury Criterion was set on a value equal to 1000[3, 7]. Value of HIC equal to 1000 represents approximately 48% skull fracture potential according to [5]. The probability of skull fracture potential tied up to the HIC value can be seen in Fig. 8.

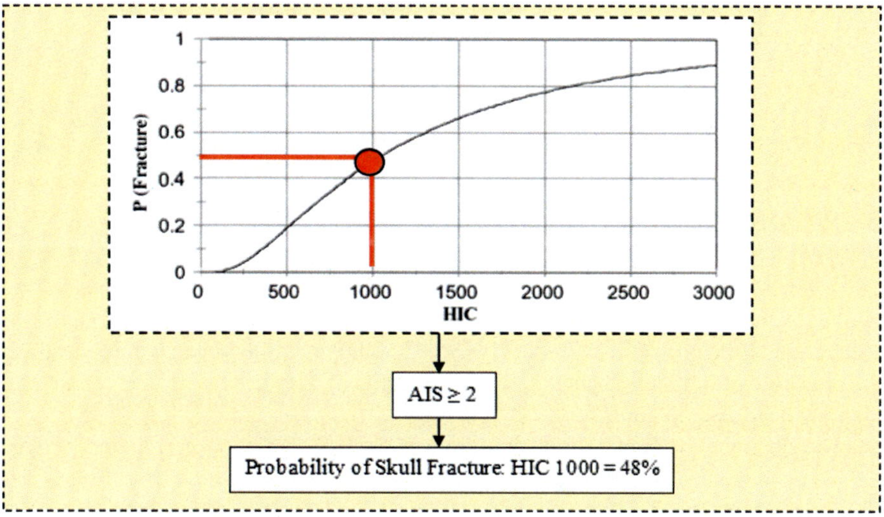

Fig. 8. Probability of skull fracture (adapted from [7]).

The HIC calculation is given by Eq. (4) and sub calculation through Eq. (5). The Eqs. (4) and (5) were adapted from [4, 5, 7].

$$\textit{Health Injury Criterion} = \max_{t_1,t_2}\left\{\left(\frac{1}{t_2 - t_1}\int_{t_2}^{t_1} a\,dt\right)^{2.5}(t_2 - t_1)\right\} \tag{4}$$

$$a = \sqrt{a_x^2 + a_y^2 + a_z^2} \tag{5}$$

The symbol a in Eq. (4) represents absolute acceleration obtained from Hybrid III 50th Percentile Male Dummy head motion. Symbols t_1 and t_2 represent the calculation interval used for obtaining the HIC value – for our case, the time interval is 15 ms. 15 ms time interval is widely used for analyzing short-duration impacts acting on the human head. The symbols a_x, a_y, and a_z in Eq. (5) represent the resultant head acceleration in the orthogonal coordinate system (see Fig. 7). X coordinate is the direction of the vehicle's motion. The acceleration components are tied up to the Hybrid III 50th Percentile Male Dummy head center of gravity – placement of the triaxial accelerometer.

The human head injury potential approach is computed for three cases of projectile weight – 0.5 kg, 1.0 kg, and 1.5 kg. The projectile's weight plays a significant role in the injury potential calculated through Blunt Criterion and Head Injury Criterion.

Mass representing the struck individual must also be carefully selected. A comparative calculation of Blunt Criterion for two separate struck object masses was conducted – one for whole struck object body mass and one for struck body part only – see Table 2.

Table 2. Specific mass values referring to the struck individual/object (Hybrid III 50th Percentile Male Dummy).

Calculation Case No	Mass in kg	Adapted from
1	4.54 (Only Head)	[5]
2	75.39 (Whole Body)	[5]

3 Results

The experimental test performance can be seen in Fig. 9. Figure 9 describes through photos taken from the high-speed camera the projectile movement from the airgun barrel, interaction of the projectile with the Hybrid III 50th Percentile Male Dummy head, and the projectile post impulse behavior of the projectile and Hybrid III 50th Percentile Male Dummy head after the impact.

The obtained HIC and Blunt Criterion results show a good correlation between the obtained results from both used calculation methods. The results from Blunt Criterion calculations are strongly dependent on the selected values of the struck individual masses and anthropometric characteristics (skull and soft tissue thickness) of the struck individual body part.

Figure 10 represents the situation when 0.5 kg stiff projectile interacts with Hybrid III 50th Percentile Male Dummy head. The projectile has different velocities (initial conditions). The mass of the struck individual/object shifts the Blunt Criterion critical value to higher projectile velocity levels – the critical projectile velocity (causing the potential injury) increases. When comparing the Blunt Criterion results to the HIC critical value of 1000, the significant result correlation lies near the critical projectile speed value of 60 km/h (if HIC is the comparator).

Figure 11 describes a case when a 1.0 kg stiff projectile interacts with Hybrid III 50th Percentile Male Dummy head. The projectile has different velocities (initial conditions).

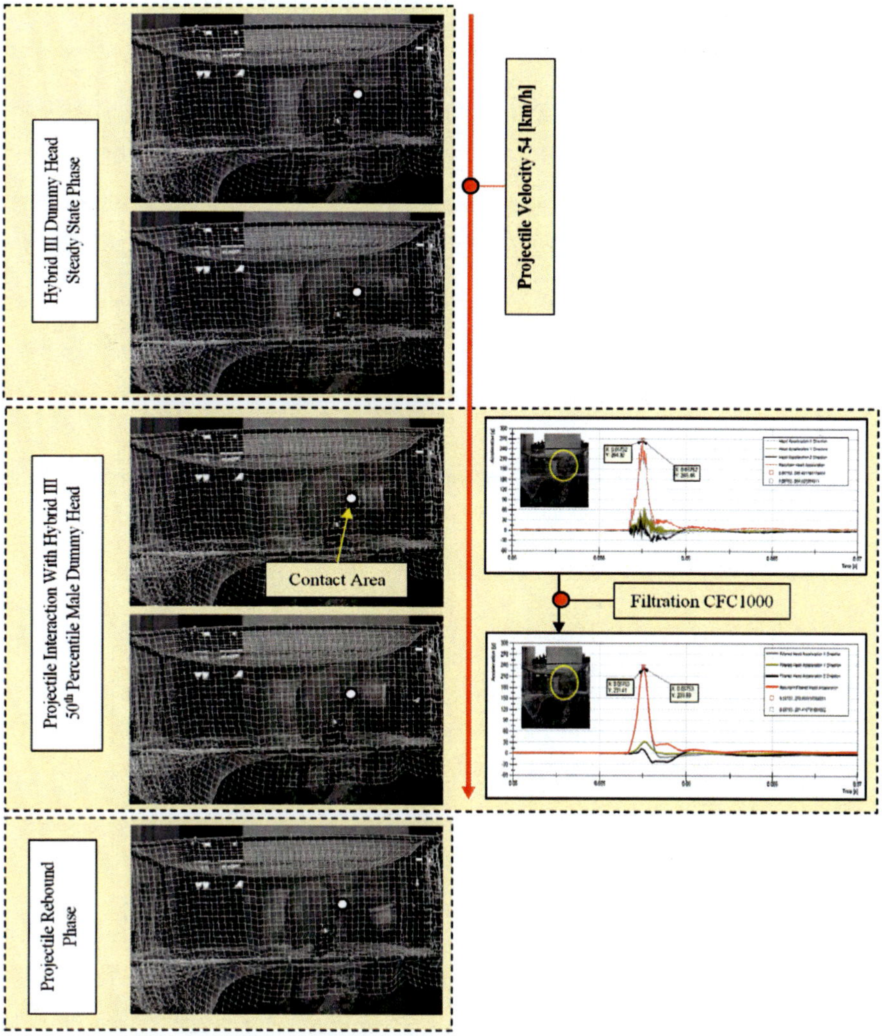

Fig. 9. Experimental process visualization – the Hybrid III 50th Percentile Male Dummy head and projectile interaction with initial impact velocity 54 km/h.

The mass of the struck individual/object shifts the Blunt Criterion critical value to higher projectile velocity levels (as for 0.5 kg projectile) – the critical projectile velocity (causing the potential injury) increases. When comparing the Blunt Criterion results to the HIC critical value of 1000 [1, 7], the significant result correlation lies near the critical projectile speed value of 45 km/h (if HIC is the comparator).

Figure 12 describes a case when 1.5 kg stiff projectile interacts with Hybrid III 50th Percentile Male Dummy head. The projectile has different velocities (initial conditions). The mass of the struck individual/object shifts the Blunt Criterion critical value to higher projectile velocity levels (as for 0.5 kg and 1.0 kg projectile) – the critical projectile

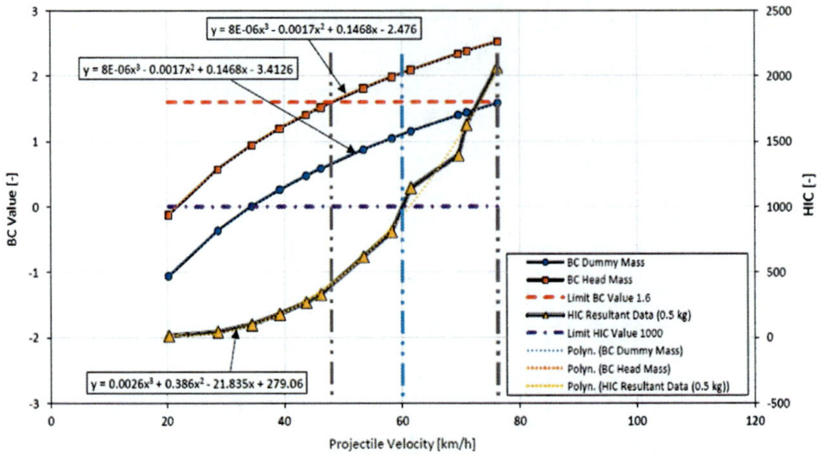

Fig. 10. Obtained Blunt Criterion and HIC values – comparison of struck individual different masses (whole-body vs. head) – projectile mass 0.5 kg.

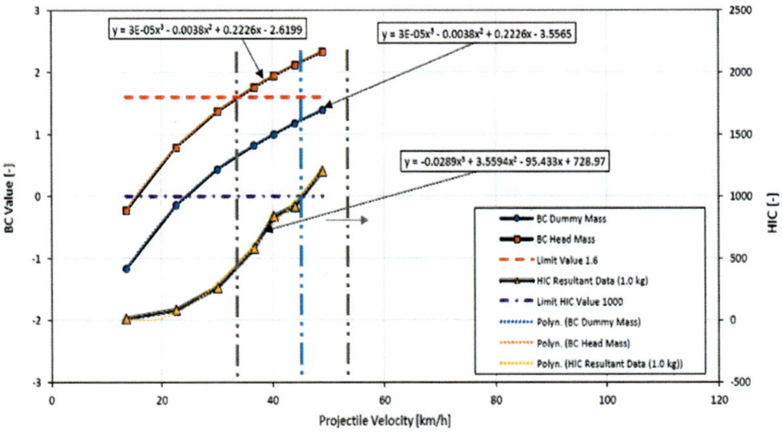

Fig. 11. Obtained Blunt Criterion and HIC values – comparison of struck individual different masses (whole-body vs. head) – projectile mass 1.0 kg.

velocity (causing the potential injury) increases. When comparing the Blunt Criterion results to the HIC critical value of 1000 [3] [7], the significant result correlation lies near the critical projectile speed value of 38 km/h (if HIC is the comparator).

Results characterizing Blunt Criterion calculation using only the head mass but different skull and soft tissue thickness can be seen in Fig. 13, 14, 15. Higher skull and soft tissue thickness values shift the Blunt Criterion critical value (1.6 [1]) to higher projectile velocities.

Figure 13 represents a situation when 0.5 kg projectile struck the Hybrid III 50th Percentile Male Dummy head. A good correlation between HIC critical value (1000 [3, 7]) and Blunt Criterion critical value (1.6) [1] is for skull and soft tissue thickness equal to

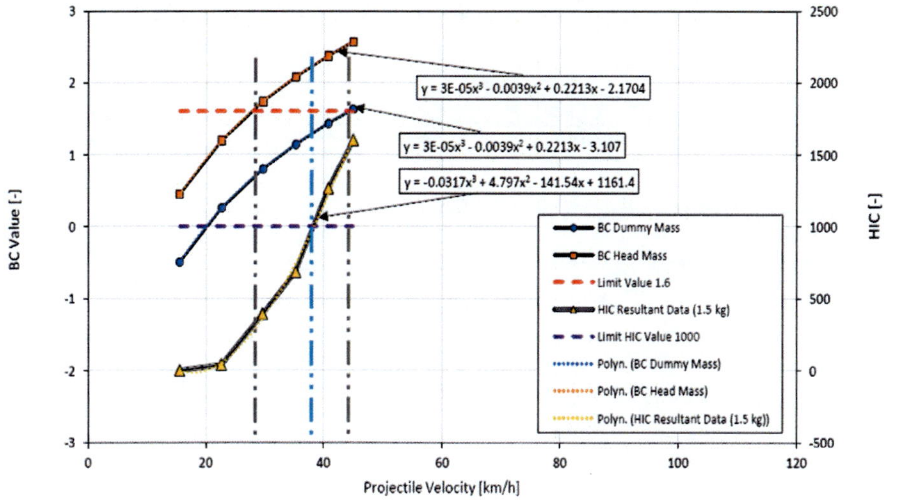

Fig. 12. Obtained Blunt Criterion and HIC values – comparison of struck individual different masses (whole-body vs. head) – projectile mass 1.5 kg.

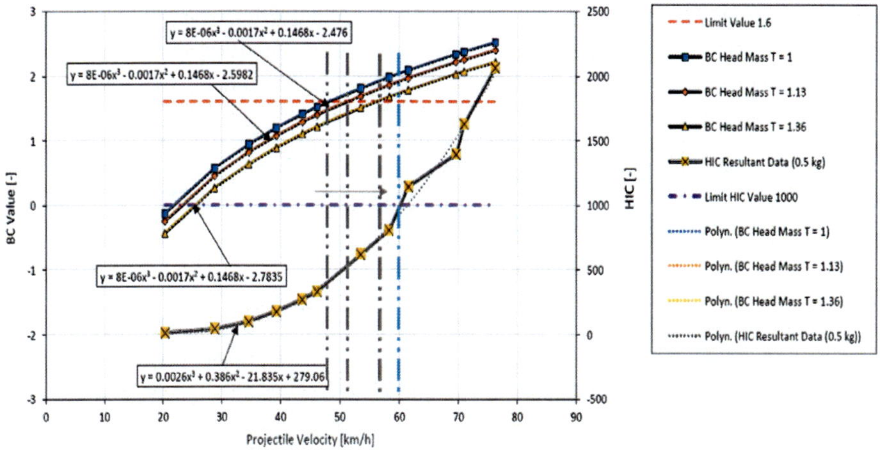

Fig. 13. Obtained Blunt Criterion and HIC values – comparison of a different skull and soft tissue thicknesses for struck human part (head) – projectile mass 0.5 kg.

1.36 cm – for this case, the critical projectile velocity is: regarding HIC circa 60 km/h; regarding Blunt Criterion circa 57 km/h.

Figure 14 represents a situation when 1.0 kg projectile strikes the Hybrid III 50th Percentile Male Dummy head. A good correlation between HIC critical value (1000 [3, 7]) and Blunt Criterion critical value (1.6 [1]) is for skull and soft tissue thickness equal to 1.36 cm – for this case, the critical projectile velocity is: regarding HIC circa 45 km/h; regarding Blunt Criterion circa 40 km/h.

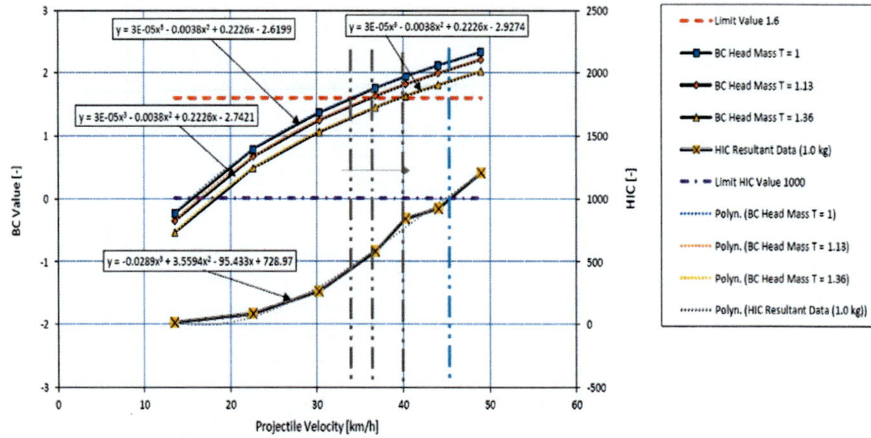

Fig. 14. Obtained Blunt Criterion and HIC values – comparison of a different skull and soft tissue thicknesses for struck human part (head) – projectile mass 1.0 kg.

Figure 15 represents a situation when 1.5 kg projectile strikes the Hybrid III 50th Percentile Male Dummy head. A good correlation between HIC critical value (1000 [3, 7]) and Blunt Criterion critical value (1.6 [1]) is for skull and soft tissue thickness equal to 1.36 cm – for this case, the critical projectile velocity is: regarding HIC circa 38 km/h; regarding Blunt Criterion circa 33 km/h.

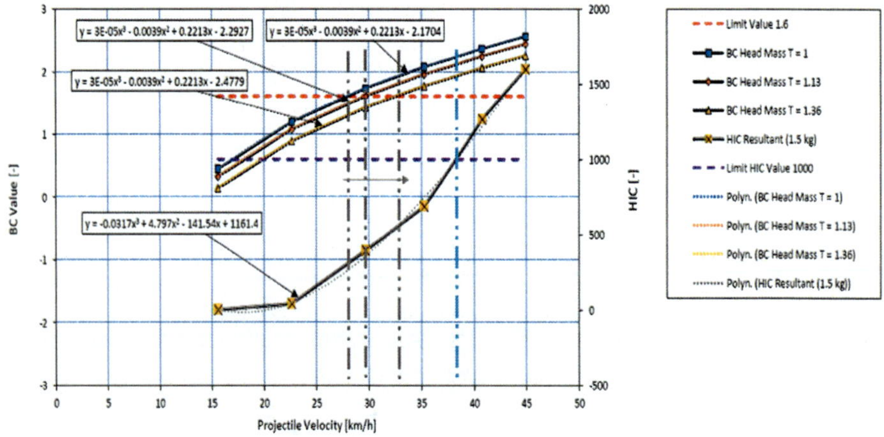

Fig. 15. Obtained Blunt Criterion and HIC values – comparison of a different skull and soft tissue thicknesses for struck human part (head) – projectile mass 1.5 kg.

HIC calculations presented in Fig. 10, 11, 12, 13, 14, 15 are used as a comparator to obtained Blunt Criterion values. The HIC values were chosen as a comparator since HIC is widely used in practice to determine the potential of head injury. Blunt Criterion is used mainly to study human body vs. projectile interactions in forensic applications focused mainly on wound ballistics.

Equations in Fig. 10, 11, 12, 13, 14, 15 describe specific functions that characterize the probable evolution of HIC or Blunt Criterion values regarding the projectile velocity values. The equations are computed in MS Excel through the trendline function. The used trendline function has a polynomial order equal to 3, matching the obtained HIC and Blunt Criterion data reasonably.

4 Discussion

The results presented in this study are mainly based on the experimental approach using Hybrid III 50th Percentile Male Dummy, which represents a realistic biomechanical model of the human body and its parts. Head responses and projectile velocities are measured, and the HIC and Blunt Criterion are calculated.

The Head Injury Criterion (HIC) calculation is based only on the assumption that one specific projectile velocity is needed to obtain the Hybrid III 50th Percentile Male Dummy head response data without repetition. Typically, three or more repetitions on the same velocity are needed to obtain relevant data (statistics). This approach was dismissed – Potential problems with calibrating the sensors on Hybrid III 50th Percentile Male Dummy due to projectile velocities applied; In practice, no repetitions of crash tests are performed to obtain statistical data.

The values of Blunt Criterion are strongly dependent on the user-defined inputs. If the inputs are not carefully selected, Blunt Criterion calculations give unsatisfactory results. Based on the obtained Blunt Criterion results compared with HIC values, the mass of the struck human body part should be selected (not the whole body). Also, the weight of the human body part plays a role in the resultant calculated data – the higher the weight is, the lower the critical projectile velocities are.

A similar statement is applicable for the skull and soft tissue thickness – the higher the combined thickness is, the lower the critical projectile velocities are.

The Blunt Criterion results relating to the skull and soft tissue thickness (combined) are spread in the range of about 6 km/h for all three solved cases (projectile mass 0.5 kg, 1.0 kg, and 1.5 kg). Blunt Criterion calculation results using higher thickness values correlate better with HIC values. However, all Blunt Criterion calculations using the different skull and soft tissue thicknesses give satisfactory results regarding the uncertainty tied up to the anthropometric human head characteristics.

The assumption can be taken that the rear part of the human head has a small portion of skin, and the realistic combined thickness of the skull and soft tissue can be approximately 1 cm or less (in obtained results – the highest difference between HIC and Blunt Criterion value). If we consider some amount of hair, the thickness can increase above 3 or 4 mm. Since the Hybrid III, 50th Percentile Male Dummy head has a constant thickness of the soft tissue and constant thickness of the "skull" approx. 1.1 cm [6], the all obtained results match very well.

If the statistical analysis (confidence interval, other) is performed for both (Blunt Criterion and HIC), the difference regarding obtained values with variable values of a thickness (skull and soft tissue) T and mass W vill be more negligible. A "pseudo" statistical analysis will be performed in the future using statistical calculation tools in the software OriginPro.

In reality, the human head's response to impact during a vehicular crash is a highly complex function that can include head interactions with different parts of the car's inner structure, airbags, and others. The study focuses only on a specific case of head and projectile interaction, which is not so complex/complicated and can be adequately analyzed.

References

1. Frank, M., et al.: Blunt Criterion trauma model for head and chest injury risk assessment of cal. 380 R and cal. 22 long blank cartridge actuated gundog retrieval devices. Foren. Sci. Int. **208**(1-3), 37–41. ISSN 0379–0738 (2011)
2. Sturdivan, L.M., Viano, D.C., Champion, H.R.: Analysis of injury criteria to assesschest and abdominal injury risks in blunt ballistic impacts. J. Trauma **56**, 651–663 (2004)
3. Mackay, M.: The Increasing Importance of the Biomechanics of Impact Trauma, Sadhana, pp. 397–408 (2007)
4. King, A.: The Biomechanics of Impact Injury Response, Mechanisms of Injury, Human Tolerance and Simulation (2017). https://doi.org/10.1007/978-3-319-49792-1
5. 78051-9901: User Manual HIII-50th FAA Male, Rev A Page 1 of 110, © 2009 First Technology Safety Systems, Inc. (2009)
6. Raymond, D., Van Ee, C., Crawford, G., Bir, C.: Tolerance of the skull to blunt ballistic temporo-parietal impact. J. Biomech. **42**(15), 2479–2485 (2009). https://doi.org/10.1016/j.jbiomech.2009.07.018. PMID: 19674749
7. Schmitt, K.-U., Niederer, P., Walz, F.H.: Trauma Biomechanics: Introduction to Accidental Injury, 5 edn. Springer, Berlin (2014). https://doi.org/10.1007/978-3-642-53920-6

Using Spherical Contact Pairs to Model the Contact Areas in the Joints of the Wrist

Paweł Ikoniak and Adam Ciszkiewicz(✉)

Faculty of Mechanical Engineering, Cracow University of Technology, Cracow, Poland
adam.ciszkiewicz@pk.edu.pl

Abstract. This study aimed to analyze the possibility of using spherical contact pairs to model the articular surfaces of the joints of the wrist. To do so, a three-dimensional surface model of the bones in the wrist was created based on medical scans from computer tomography. Twenty-two pairs of surfaces, which represented the contact areas, were selected and cut out of the model. After that, these pairs were imported into custom software in Python, which allowed for numerical sphere fitting using optimization. The obtained results were then analyzed in terms of the validity of the obtained contact pairs, whether they formed actual ball-and-socket joints and based on fit quality. To verify the results obtained from the first model, a second set of meshes was obtained. The obtained results showed that for most of the joints in the wrist, the spherical approximation might be a viable choice. In some of the cases, the returned radii of the spheres were unusually large, which could mean that different contact pairs, such as the sphere-plane type might be more accurate.

Keywords: Bone contact · Sphere fitting · Optimization

1 Introduction

The wrist contains eight bones, which can be subdivided into two sets: proximal and distal. It is a bridge between the five digits and the radius/ulna forming the elbow joint. In total, the wrist connects fifteen bones in twenty-two joints, which makes it one of the more complicated structures in the human body. Because of this complexity, many different approaches to wrist modeling are available in literature. Probably the most common approach is to substitute the wrist with only single or dual constraints, usually in the form of hinge joints [1–7]. Nevertheless, such an approach can be limiting, especially when considering the contact behavior in the joint. This problem is usually addressed with the rigid body spring method [8] or the finite element method [9–14], in which all joints and their cartilage are represented. However, due to the sheer number of joints, their sizing, etc., obtaining viable meshes for use in contact analysis can be difficult and may require manual adjustment of geometry, as in [15]. The method itself is also costly numerically, especially when considering complex, dynamic models. In the literature, a different approach to contact modeling in joints has also been studied. Deformable contact pairs have been proposed to substitute the cartilage in the knee [16, 17], the hip

A. Hadamus et al. (Eds.): BIOMECHANICS 2021, LNNS 328, pp. 71–79, 2022.
https://doi.org/10.1007/978-3-030-86297-8_7

[18], and the ankle [19, 20]. This method allows for a more accurate description of the bone contact behavior of the joint than the regular spherical/cylindrical constraints and is also fairly inexpensive when compared to the finite element approach. To the best of our knowledge, it has never been applied to the joints of the wrist.

The aim of this research was to perform a preliminary analysis on whether the contact areas in all twenty-two subjoints in the wrist could be simplified using deformable spherical contact pairs. This study was limited to a geometrical analysis of the bone profiles. The method and the obtained results were presented in the following paragraphs.

2 Method

2.1 Obtaining Geometrical Models of the Joints in the Wrist

In this study, the geometrical models of wrist's joints were based on computer tomography (CT) scans. The procedure was performed on two separate CT datasets from *embodi3d.com*, available under Creative-Commons By Attribution license:

- filename: ct wrist 1.0.0 by *Mahipal* at https://www.embodi3d.com/files/file/38758-ct-wrist/, under *CC – Attribution license* and accessed: 13.10.2020,
- filename: *my wrist 1.0.0* by *rajwardhan19* at https://www.embodi3d.com/files/file/41972-wrist/, under *CC – Attribution* license and accessed: 29.04.2021.

Nevertheless, due to the quality of the second scan, only two contact pairs were created from it. Therefore, the first scan was used to check all subjoints, while the second one functioned as a partial verification for the results from the first.

In the first step, the image segmentation was carried out. To do this, CT scans were imported into *3D Slicer*. Then, with the *Threshold* function, the bones were separated from other tissues. After that, *3D Slicer* was used to create a surface mesh of the bones in the wrist with a marching cubes approach – see Fig. 1. The meshes were cleaned out and smoothed in *Meshmixer*.

To model joint connections, twenty-two pairs of contact surfaces were cut from the bone models in *Meshmixer* – see Fig. 2. The contact patches were selected taking into account the anatomical placement of the joint's cartilage. After a preliminary analysis of the shape of the contact surfaces, it was devised that the subjoints resemble a ball-and-socket joints, therefore, simple contact pairs of two spheres could likely be used to substitute the cartilage in them. A simple script was written in *Python* to fit spheres to the extracted contact surfaces. The software used the *STL* (*Standard Tessellation Language*) file format for the meshes, while the sphere fitting was performed with an optimizational approach based on [21].The script also allowed for visualization of the obtained contact pairs (with and without the original meshes) in *mayavi* and *matplotlib*.

Note that in this case the contact patches representing cartilage were based on the meshes of the bones obtained from CT scans, which is common in biomechanics [9, 22]. Nevertheless, the proper cartilage thickness could be set after sphere fitting by adjusting the radius of the spheres using experimental data [23–25].

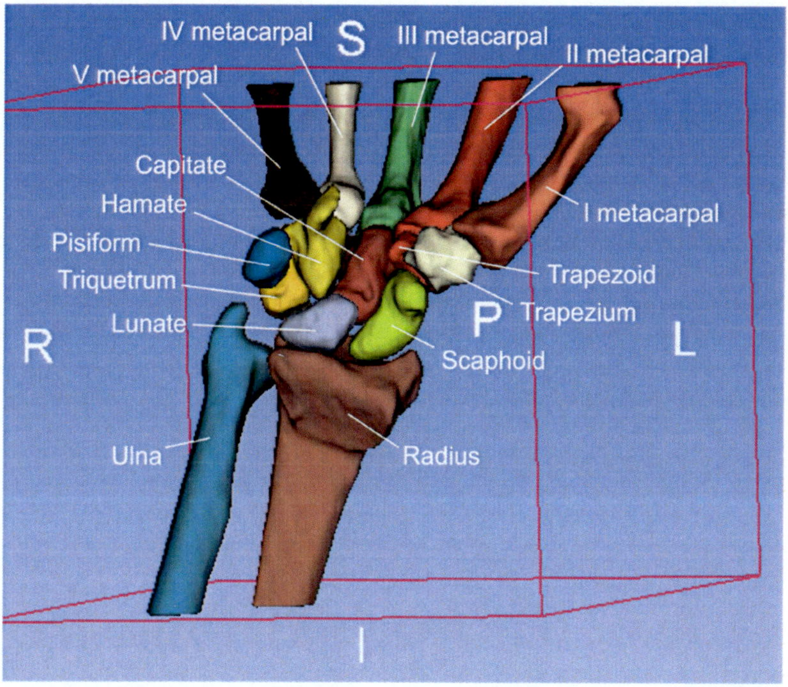

Fig. 1. A sample of the obtained three-dimensional meshes as seen in *3D Slicer*.

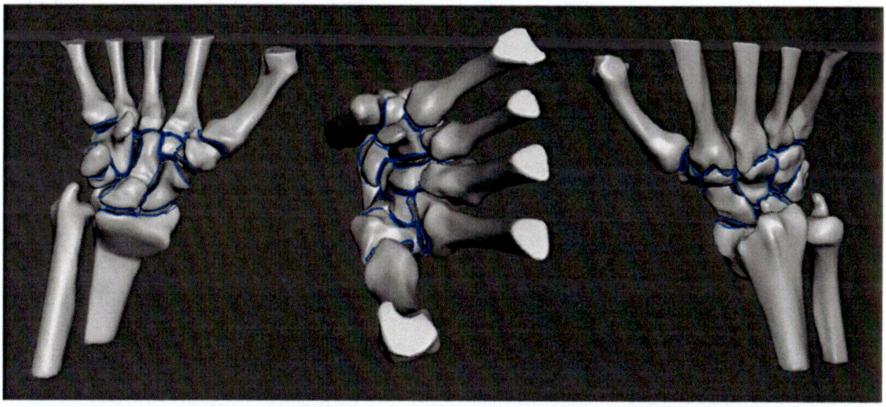

Fig. 2. The articular surfaces of the twenty-two joints as seen in *Meshmixer*.

2.2 Checking the Quality of the Contact Pairs

The quality check of the obtained contact pairs was performed in two ways. Firstly, we employed standard deviation to assess the quality of the sphere fitting procedure for each sphere in all contact pairs. Then, visual representations of the contact pairs were used to check whether the pairs were valid in an anatomical sense, i.e., the pairs resembled proper ball-and-socket joints. Based on the initial assessment of the results, we distinguished three different cases:

- a valid contact pair – the anatomically smaller sphere (ball) was contained within the anatomically larger sphere (socket); 1.0 point
- a partially invalid contact pair – the two spheres intersected (due to curvature of the contact patch or numerical problems during fitting), it might be possible to manually adjust it into a valid contact pair, with, for instance, minor changes to the sphere centers; 0.5 points,
- an invalid contact pair – the two spheres were convex and did not form a ball-and-socket joint (due to curvature of the contact patch or numerical problems during fitting), which might suggest that a different type of contact pair should be used to model the joint; 0.0 points.

These cases were visualized in Fig. 3.

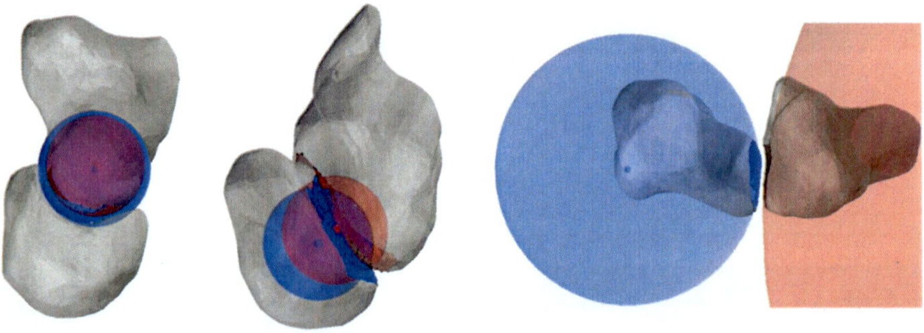

Fig. 3. The three variants of the contact pairs, which occurred in the datasets, from the left: valid, partially-invalid, invalid.

3 Results and Discussion

This part of the paper was subdivided into two sections. In the first one, we presented the results obtained for the twenty-two contact pairs from the first CT dataset obtained with two sphere-fitting procedures. In the second section, the selected contact pairs were analyzed in detail with regards to how they differ between the first and the second CT dataset.

3.1 Twenty-Two Contact Pairs in the First Model

The results obtained from the first dataset were presented in Table 1. Valid contact pairs were obtained in almost 60% of the studied subjoints, while invalid contact pairs constituted only 23% of the cases. Notably, the proper pairs were observed mostly for the joints occurring in the proximal part of the wrist, involving the scaphoid and lunate bones. The worst results were obtained for the joints, which connect the wrist to the digits. In these cases, completely invalid pairs were common, which suggests that the sphere-sphere simplification might not be adequate to fully substitute the cartilage in these connections.

The results also contained several instances, in which the fitting procedure failed to deliver plausible results. This was reflected by unusually large values of radii for some pairs. Although we were unable to fully ascertain the cause of these results, we suspect that it was a failure of the search procedure. Nevertheless, a very large value of radius might also suggest that, rather than sphere-sphere, plane-sphere contact should be employed. This issue requires further study.

Table 1. Details regarding the contact pairs obtained from the first dataset, where: italic font signifies partially-invalid pairs and bold font marks invalid contact pairs.

	First bone	Second bone	r [mm]	R [mm]	pts
1	Scaphoid	Radius	12.16	20.17	1.0
2	Scaphoid	Trapezoid	5.28	24.47	1.0
3	Scaphoid	Trapezium	7.09	14.70	1.0
4	Scaphoid	Lunate	10.16	90.38	1.0
5	Scaphoid	Capitate	7.40	10.24	1.0
6	*Lunate*	*Radius*	*17.89*	*18.93*	*0.5*
7	Lunate	Capitate	6.15	7.00	1.0
8	Lunate	Hamate	6.97	24.69	1.0
9	Lunate	Triquetrum	11.27	51.25	1.0
10	Triquetrum	Pisiform	9.55	15.37	1.0
11	Triquetrum	Hamate	42139.00	44537.00	1.0
12	*Trapezium*	*I metacarpal bone*	*11.21*	*12.68*	*0.5*
13	**Trapezium**	**II metacarpal bone**	**17.90**	**64.62**	**0.0**
14	**Trapezium**	**Trapezium**	**22.27**	**8671.00**	**0.0**
15	**Trapezium**	**II metacarpal bone**	**67.60**	**52782.00**	**0.0**
16	Trapezium	Capitate	12.80	15.68	1.0
17	*Capitate*	*II metacarpal bone*	*14.70*	*25.55*	*0.5*
18	**Capitate**	**III metacarpal bone**	**16.62**	**316.00**	**0.0**
19	*Capitate*	*IV metacarpal bone*	*3.85*	*139164.00*	*0.5*
20	Capitate	Hamate	24.39	41.66	1.0
21	**Hamate**	**IV metacarpal bone**	**20.39**	**852.00**	**0.0**
22	Hamate	V metacarpal bone	9.84	13.59	1.0

3.2 A Preliminary Verification of Three Contact Pairs on the Second Dataset

As mentioned in the method section, two three-dimensional models were created based on two CT datasets. Due to low quality, only two contact pairs were obtained from the second dataset. In this section, the two pairs from the second set were verified in detail against the corresponding pairs from the first dataset. These pairs were between:

- the scaphoid and radius,
- the capitate and lunate.

As seen in Fig. 4, the first of the contact pairs between the scaphoid and the radius was valid for both the first and the second model. Naturally, the sizing and the overall geometry differed between the models. Nevertheless, in both cases, the connection was a proper ball-and-socket joint – the spheres did not intersect and the anatomically larger one formed a concave socket for the smaller, convex ball.

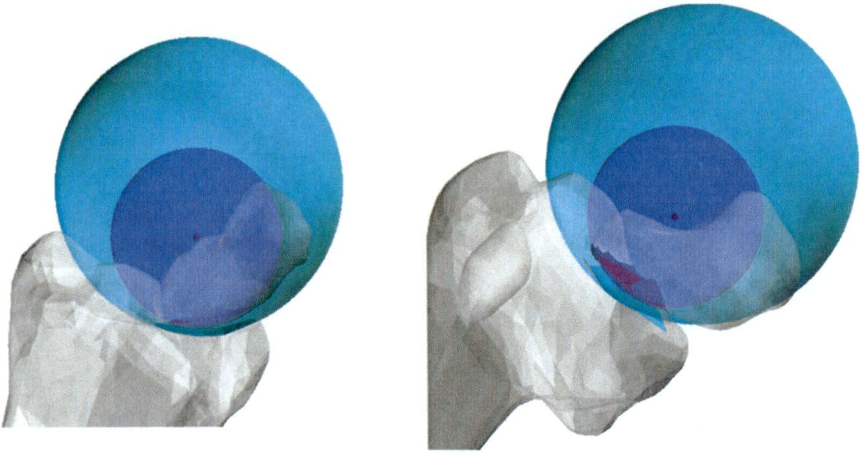

Fig. 4. The contact pair between the scaphoid and the radius bones for the first (on the left) and the second (on the right) dataset.

The details regarding the geometry and the quality of the sphere fit for the contact pair were presented in Table 2. In both models, the spheres in the contact pair showed a relatively good fit to the meshes with a standard deviation under 0.45 mm. Furthermore, the mean absolute distance of the mesh points from the sphere was no larger than 0.38 mm. When compared to the radius of the smallest of the spheres – 12.40 mm – the result could be considered good.

Similar results were obtained for the pair between capitate and lunate, which was presented in Fig. 5. Both datasets returned proper and valid contact pairs in this case. The spheres formed a ball-and-socket joint and did not intersect. Regarding the fitting quality, as seen in Table 3, the spheres obtained for the second model were slightly larger than in the first one – as in the scaphoid and radius case. Again, every sphere showed

Table 2. The details regarding the scaphoid and radius contact pair in the two models, where r and σ stand for *radius, standard deviation.*

	The scaphoid		*The radius*	
Model	r [mm]	σ [mm]	r [mm]	σ [mm]
1	*12.40*	*0.19*	*11.40*	*0.20*
2	*16.18*	*0.27*	*20.21*	*0.44*

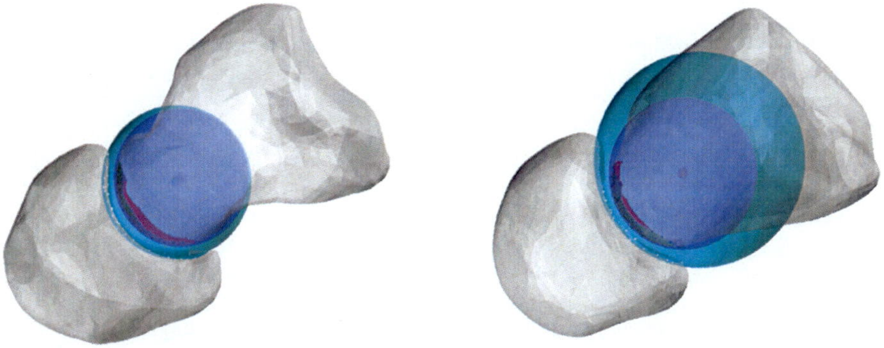

Fig. 5. The contact pair between capitate and lunate bones for the first (on the left) and the second (on the right) dataset.

Table 3. The details regarding the capiate and lunate contact pair in the two models, where r and σ stand for *radius, standard deviation.*

	The capiate		*The lunate*	
Model	r [mm]	σ [mm]	r [mm]	σ [mm]
1	*6.16*	*0.17*	*6.71*	*0.26*
2	*6.90*	*0.13*	*9.08*	*0.39*

a good fit to the contact mesh, with the mean absolute distance from a sphere never exceeding 0.29 mm for the sphere with the radius of 9.08 mm.

4 Conclusion

The paper presented an analysis of the contact areas of the joints available in the wrist. Twenty-two unique subjoints were analyzed to test whether their geometry allowed for describing their articular surfaces with simple sphere-sphere contact pairs. The contact pairs were obtained through a sphere fitting procedure with input based on two segmented CT datasets.

Our results indicated that in almost 60% of the studied joints, the articular surfaces could be simplified to a sphere-sphere contact pair. These cases corresponded to the proximal part of the wrist. Some of these results were verified with the second model. On the other hand, for 23% of the joints in the distal part of the wrist, we were unable to obtain valid contact pairs. In some cases, the fitting procedure returned unusually high values of radii for the spheres. Based on our current results, we were unable to ascertain the source of this phenomenon. Nevertheless, we suspect that it might indicate that different contact pairs, such as the plane-sphere type, should be employed to model these joints. This issue requires further study.

References

1. Holzbaur, K.R.S., Murray, W.M., Delp, S.L.: A model of the upper extremity for simulating musculoskeletal surgery and analyzing neuromuscular control. Ann. Biomed. Eng. **33**, 829–840 (2005). https://doi.org/10.1007/s10439-005-3320-7
2. Qin, J., Lee, D., Li, Z., Chen, H., Dennerlein, J.T.: Estimating in vivo passive forces of the index finger muscles: exploring model parameters. J. Biomech. **43**, 1358–1363 (2010). https://doi.org/10.1016/j.jbiomech.2010.01.014
3. Lemay, M.A., Crago, P.E.: A dynamic model for simulating movements of the elbow, forearm, and wrist. J. Biomech. **29**, 1319–1330 (1996). https://doi.org/10.1016/0021-9290(96)00026-7
4. Laitenberger, M., Raison, M., Périé, D., Begon, M.: Refinement of the upper limb joint kinematics and dynamics using a subject-specific closed-loop forearm model. Multibody Sys.Dyn. **33**(4), 413–438 (2014). https://doi.org/10.1007/s11044-014-9421-z
5. Weinberg, A.M., Pietsch, I.T., Helm, M.B., Hesselbach, J., Tscherne, H.: A new kinematic model of pro- and supination of the human forearm. J. Biomech. **33**, 487–491 (2000). https://doi.org/10.1016/S0021-9290(99)00195-5
6. Duprey, S., Naaim, A., Moissenet, F., Begon, M., Chèze, L.: Kinematic models of the upper limb joints for multibody kinematics optimisation: an overview. J. Biomech. **62**, 87–94 (2017). https://doi.org/10.1016/j.jbiomech.2016.12.005
7. Nojiri, K., Matsunaga, N., Kawaji, S.: Modeling of pro-supination for forearm skeleton based on MRI. IFAC (2008). https://doi.org/10.3182/20080706-5-kr-1001.02500
8. Fischli, S., Sellens, R.W., Beek, M., Pichora, D.R.: Simulation of extension, radial and ulnar deviation of the wrist with a rigid body spring model. J. Biomech. **42**, 1363–1366 (2009). https://doi.org/10.1016/j.jbiomech.2009.03.008
9. Bajuri, M.N., Kadir, M.R.A., Raman, M.M., Kamarul, T.: Mechanical and functional assessment of the wrist affected by rheumatoid arthritis: a finite element analysis. Med. Eng. Phys. **34**, 1294–1302 (2012). https://doi.org/10.1016/j.medengphy.2011.12.020
10. Varga, P., Schefzig, P., Unger, E., Mayr, W., Zysset, P.K., Erhart, J.: Finite element based estimation of contact areas and pressures of the human scaphoid in various functional positions of the hand. J. Biomech. **46**, 984–990 (2013). https://doi.org/10.1016/j.jbiomech.2012.11.053
11. Varga, P., Zysset, P.K., Schefzig, P., Unger, E., Mayr, W., Erhart, J.: A finite element analysis of two novel screw designs for scaphoid waist fractures. Med. Eng. Phys. **38**, 131–139 (2016). https://doi.org/10.1016/j.medengphy.2015.11.006
12. Gislason, M.K., Nash, D.H., Stansfield, B.: In vivo contact stresses at the radiocarpal joint using a finite element method of the complete wrist joint. J. Biomech. **41**, S147 (2008). https://doi.org/10.1016/s0021-9290(08)70147-7
13. Akhbari, B., et al.: In vivo articular contact pattern of a total wrist arthroplasty design. J. Biomech. **121**, 110420 (2021). https://doi.org/10.1016/j.jbiomech.2021.110420
14. Camus, E.J., Aimar, A., Van Overstraeten, L., Schuind, F., Innocenti, B.: Lunate loads following different osteotomies used to treat Kienböck's disease: a 3D finite element analysis. Clin. Biomech. **78**, 105090 (2020). https://doi.org/10.1016/j.clinbiomech.2020.105090

15. Gíslason, M.K., Stansfield, B., Nash, D.H.: Finite element model creation and stability considerations of complex biological articulation: the human wrist joint. Med. Eng. Phys. **32**, 523–531 (2010). https://doi.org/10.1016/j.medengphy.2010.02.015
16. Machado, M., et al.: Development of a planar multibody model of the human knee joint. Nonlinear Dyn. **60**, 459–478 (2009). https://doi.org/10.1007/s11071-009-9608-7
17. Machado, M., Flores, P., Ambrosio, J., Completo, A.: Influence of the contact model on the dynamic response of the human knee joint. Proc. Inst. Mech. Eng. Part K J. Multi-body Dyn. **225**, 344–358 (2011). https://doi.org/10.1177/1464419311413988
18. Askari, E., Flores, P., Dabirrahmani, D., Appleyard, R.: Nonlinear vibration and dynamics of ceramic on ceramic artificial hip joints: a spatial multibody modelling. Nonlinear Dyn. **76**(2), 1365–1377 (2014). https://doi.org/10.1007/s11071-013-1215-y
19. Borucka, A., Ciszkiewicz, A.: A planar model of an ankle joint with optimized material parameters and Hertzian contact pairs. Materials (Basel) **12**, 2621 (2019)
20. Ciszkiewicz, A.: Analyzing uncertainty of an ankle joint model with genetic algorithm. Materials (Basel) **13**, 1175 (2020). https://doi.org/10.3390/ma13051175
21. Various: Least squares circle. https://scipy-cookbook.readthedocs.io/items/Least_Squares_Circle.html
22. Wei, F., Fong, D.T.P., Chan, K.M., Haut, R.C.: Estimation of ligament strains and joint moments in the ankle during a supination sprain injury. Comput. Methods Biomech. Biomed. Engin. **18**, 243–248 (2015). https://doi.org/10.1080/10255842.2013.792809
23. Draper, C.E., et al.: Is cartilage thickness different in young subjects with and without patellofemoral pain? Osteoarthr. Cartil. **14**, 931–937 (2006). https://doi.org/10.1016/j.joca.2006.03.006
24. Shepherd, D.E.T., Seedhom, B.B.: Thickness of human articular cartilage in joints of the lower limb. Ann. Rheum. Dis. **58**, 27–34 (1999). https://doi.org/10.1136/ard.58.1.27
25. Foumani, M., Strackee, S.D., Van De Giessen, M., Jonges, R., Blankevoort, L., Streekstra, G.J.: In-vivo dynamic and static three-dimensional joint space distance maps for assessment of cartilage thickness in the radiocarpal joint. Clin. Biomech. **28**, 151–156 (2013). https://doi.org/10.1016/j.clinbiomech.2012.11.005

3D Bioreactors for Cell Culture: Fluid Dynamics Aspects

Natalia Kizilova[1,2](✉) and Jacek Rokicki (Deceased)[1]

[1] Warsaw University of Technology,
Institute of Aeronautics and Applied Mechanics, Warsaw, Poland
[2] V.N. Karazin, Kharkov National University, Kharkiv, Ukraine

Abstract. Reconstructive therapy is essential in functionality restoration of the tissues impaired by congenital disorders, degenerative diseases and trauma that needs authentic cells for transplantation and tissue engineering. Petri dish and Cell Culture Flasks produce the cells which properties were changed by the contacts between the cells and the walls of the vessel. A bioreactor for tissue engineering applications should: (i) facilitate uniform cell distribution; (ii) provide and maintain the physiological requirements of the cell (e.g., nutrients, oxygen, growth factors); (iii) increase mass transport by diffusion and convection using mixing systems of culture medium; (iv) expose the cells to vital physical stimuli; and (v) enable reproducibility, control, monitoring and automation. Besides, bioreactors should present a simple reliable design preventing possible stagnation and allowing an easy access to the engineered tissue if any problem arises in the reactor during the operational period. In this paper the state-of-the-art review on different types of the reactors existed in the market, and their benefits is presented. The review is mostly concentrated on the fluid dynamics aspects of 3D dynamic cell culture technologies.

Keywords: Tissue engineering · Cell culture · Fluid mechanics · Mechanoelectric stimulation · Computational fluid dynamics

1 Introduction

Regenerative medicine is the future of health care, and better understanding the relationships between cells and their environment for the cell culture and tissue engineering is essential for novel biotechnologies [1, 2]. Reconstructive therapies are needed to restore functionality of tissues impaired by congenital disorders, degenerative diseases and trauma [3]. Human mesenchymal stem cells (MSCs) derived from different adult tissues have extensively been used in tissue engineering studies and showed encouraging results in preclinical models of tissue healing and regeneration [4–6]. Novel regenerative technologies also use organoids, which are 3D miniature versions of human organs produced from stem cells derived from either individual patient or healthy volunteers *in vitro* followed by recapitulation of the actual organ [7]. Organoid technology is also promising for preclinical drug testing and treatment modulation as individual patient's genetic makeup.

A. Hadamus et al. (Eds.): BIOMECHANICS 2021, LNNS 328, pp. 80–99, 2022.
https://doi.org/10.1007/978-3-030-86297-8_8

New tissue engineered products cannot be developed in a commercial setting without a solid foundation of how these cells behave in their native tissue. Since the life cycles of cells, their function and interaction are regulated by the cellular microenvironment [8], a comprehensive control over the mechanical environment in the cell culture bioreactors is essential.

Petri dish and Cell Culture Flasks are the conventional designs for cell culture and tissue engineering. 3D structures can also be grown in them as a set of monolayers of adherent cells grown on flat and rigid 2D substrates. In spite of the 2D engineered structures, live tissues in our bodies are composed of cells located inside a complex 3D extracellular matrix (ECM). The cells communicate with both ECM and neighboring cells in 3D space through biochemical, mechanical and electric signals that determines tissue/organ homeostasis. Development of essential 2D structural organization and connectivity can limit/diminish such vital properties as cellular morphology, viability, proliferation, differentiation, and gene and protein expression, response to stimuli, drug metabolism, and general cell function [8]. Importance of the electric interactions between the cells and extracellular matrix is usually missed in the tissue bioreactors and tissue engineering technologies.

Bioreactors were initially developed to allow the high-mass culture of cells used for applications in diverse areas, including fermentation, wastewater treatment and purification, food processing and drug production [9]. Many of the principles established by these applications have recently been adapted for tissue engineering purposes. A bioreactor for tissue engineering applications should: (i) facilitate uniform cell distribution; (ii) provide and maintain the physiological requirements of the cell (e.g., nutrients, oxygen, growth factors); (iii) increase mass transport both by diffusion and convection using mixing systems of culture medium; (iv) expose cells to physical stimuli; and (v) enable reproducibility, control, monitoring and automation (Sladkova 2014). Besides, bioreactors should present a simple reliable design in order to prevent contamination and allow quick access to the engineered tissue if any problem arises in the system during the operational period (e.g., fluid leakage and flow obstruction). The best design must resemble the natural microenvironment for the cultures cells [10] and be automated [11].

2 Cell Culture at Static vs Dynamic Conditions

From the end of the XIX century the leading tool for cell culture in microbiology, eukaryotic and plant cell culture, regenerative biology and tissue engineering was Petri dish (Petri plate or cell-culture dish), mainly due to its convenience and simplicity. The cells grow at static conditions in glass or plastic (polystyrene) Petri dish partially filled with warm substrate containing agar/fluid with specific components like minerals, nutrients, carbohydrates, amino acids, antibiotics, blood, dyes and markers. The growing cells could be in contact with other cells, substrate, plate walls or air that promotes their polarization and, thus, influences their individual shape, biophysical and physiological properties, and fate. For proliferation cells must have no contact with walls. Static systems have also disadvantages in the mass transport of nutrients and oxygen into 3D constructs.

In natural in vivo conditions the growing and developing cells are exposed to different oscillating external forces. The cells are sensitive to mechanical stresses, and their biomechanical properties got pathologically changed by improper mechanical environment, which the cells can adapt to and control [12, 13]. Periodical blood pumping through the blood vessels, contraction of the skeletal and numerous smooth muscle cells in the walls of hollow organs, blood vessels and other ducts produce permanent oscillatory state of the extracellular matrix that is the driving force for the interstitial flows. Therefore, the external forces produce oscillating pressures and viscous shear stresses that determine normal cell development and interaction in the natural conditions. The interstitial flows also help to transport nutrients, gases, wastes, key signaling proteins and other important components throughout the tissue which is essential for tissue maintenance and pathobiology [14]. The proteins, polysaccharides and other molecules are too large to be delivered from the capillaries to the cells by diffusion only which is unreliable when the interstitial liquids are in gel phase. Slow interstitial flows essentially enhance the extracellular transport and tissue metabolism. Nature has addressed this problem in native bone by establishing a complex lacunocanalicular network within which a nutrient-rich fluid permanently circulates feeding the bone cells located in the lacunas.

Some physical parameters of the interstitial flows are given in Table 1 and the Darcy flow through the blood vessel walls and tissues has been considered as the main mechanism of cell nutrition and maintenance [14]. Some other mechanisms like electrokinetic phenomena (electrophoresis, diffusiophoresis, electroosmosis, stream potential and capillary osmosis), specific adhesion and molecular motors are also essential for the intercellular interaction and tissue development.

Dynamical (moving) bioreactors provide a homogeneous distribution of nutrients in the bioreactor chamber by dynamical fluid movement/mixing. Bioreactors for tissue engineering applications can be broadly classified in few main categories, including spinner flasks, stirred systems, rotating wall, rotating bed, waving bed, flow per- fusion bioreactors (FPB) and compression systems.

Pulsatile pressure and flow bioreactor systems are promising for culture of cardiovascular tissues which are subjected to permanent stresses in natural conditions. FPB also provide better penetration of the nutrition, oxygen, growth hormones and other components into the core of the 3D cell colony. Those bioreactors significantly improve core cell activity and density compared to the cultures grown at static conditions. The operating conditions must satisfy the physiological mechanical environment of the cells/tissues to reduce/eliminate the mechanical overstimulation effects. The applied pressures and perfusion rates enhance cell seeding efficiency and uniformity in the mass and vital quality of the cells and engineered tissues [15–17].

The shear forces are essential for fibroblasts, vascular epithelium, muscle and some other cells that need shear forces for their proper elongation and pattern formation. For the bone cells the pump speed was set from 3.5 to 200 revolutions/min for perfusion rate 0.26–14.8 mL/min, shear stress tau = 0.01–10 dynes/cm^2 due to physiologic *in vivo* $\tau = 8$–30 dynes/cm^2 range for osteocytes [18].

Cell growth in FPBs is preferable due to:

- Cells grow better under dynamic culture conditions;
- Cells have almost no contact with neighbors and will not be polarized;
- There is continuous cycling of nutrients, hormones, and permanent removal of metabolic wastes;
- The shear force produced by the flowing medium act as a mechanical stimuli signal that further promotes stem cell differentiation toward certain cell lineages.

Necessary conditions for cell growth are O_2 supply 1–9%; CO_2 supply 5–7.5% for maintaining a proper pH = 7.2–7.5; constant temperature T = 37°C; high humidity of the air/gases over the liquid surface.

Several types of FPBs have been designed

1) Bioreactors with perfusion and electrical stimulation:

   - Cardiac tissues (H. Park, MIT);
   - Embryoid bodies (Figallo andGerecht, MIT; Elvassorre, Padova);

2) Bioreactors with perfusion and mechanical loading:

   - Bone and cartilage engineering (Marolt, MIT);
   - Osteochondral plugs (Grayson and Chao, Columbia);
   - Invertebral disc (Kandel, Mt. Sinai);
   - Vasculogenesis (Kang, Tufts);
   - Vasculogenesis with human embryonic stem sells (Gerecht, MIT);
   - Blood vessels (Lovett, Tufts);
   - Human Umbilical Vein Endothelial Cells;

3) Microfluidic bioreactors allowing imaging and screening studies:

   - Microbioreactor arrays (Elvassorre, Padova);
   - Gradient bioreactors (Moon, U. Washington, Cimetta, Columbia);
   - Nanoliter scale microbioreactor array;
   - multi-shear device for investigating the effects of low fluid-induced stresses on cells.

Design and operating conditions of a FPB are preferably determined by CFD computations (AnSys Fluent) on 3D design (SolidWorks) with detailed numerical estimations of the stresses, fluid flows and cell deformations in the system with coupled chemical, electric and magnetic effects (AnSys Multiphysics) [19–22].

3 2D vs 3D Bioreactors for Cell Culture

The Petri dish provides an example of 2D cell culture. New research indicates that everything we have learned in 2D cannot be translated in the 3D. As it was stated in a recent study on behavior of the cancer cells grown in the Petri dish [23] "much of how cells behaved in a Petri dish was an artifact of the 2-D environment. The cells moved as they did not because that's how motility works in cancer cells but because the cells were in a dish. Put them in a 3D environment and everything changed". Fundamental differences between 2D and 3D cell cultures systems include cell attach, spread and grow, their morphology, proliferation, differentiation, gene and protein expression.

The cells cultured in the 2D bioreactors on the flat surface of a substrate possess non natural flattened shape due to cell adhesion to the surface. 3D cultures exhibit a higher degree of structural complexity and homeostasis, which is analogous to tissues and organs. Main advantages of 2D cell cultures are easier environmental control, cell observation, measurement and eventual manipulation in comparison to 3D cultures. However, limitations such as inability to depict traits exhibited by *in vivo* systems, decreased compatibility with *in vivo* systems, increased drug sensitivity and exposed surface hamper their use in both clinical and fundamental investigations. Advantages of 3D cell cultures are proximity of different cells at all sides, behavior akin to *in vivo* conditions, as well as more accurate representation of cytoskeleton and intracellular architecture. These types of cultures can also be used as efficient simulators of tumor characteristics such as dormancy, hypoxia and anti-apoptotic behavior. In 2D culture, a monolayer of cells is in continuous contact with culture medium, and simple diffusion is sufficient to maintain cell viability, while in the 3D constructs we need chemo transportation in bioreactors.

Several alternative culture systems like gel matrix [24–26] and sponge matrix [27] cultures have been designed. Nevertheless, they present a major disadvantage: cell densities are limited by diffusion of oxygen, nutrients and waste. Diffusional transport limitations for oxygen and other essential nutrients, as well as culture dependent alterations in gene expression are potential drawbacks of the utilization of 3D cultures. Furthermore, some 3D cultures created from specific tissues (for example, basement membrane extracts) can contain undesirable components like viruses or growth factors.

Early attempts at 3D cell culture utilized explant tissue cultures, which are the cultures of small pieces of tissue surgically removed from tissue/organ of animal/human [28]. More recent attempts at 3D modeling have included scaffold-based systems, scaffold-free systems, trans wells and micro fluidics [29, 30]. The 3d liver structure can be restored from a series of slices of 2d modified polyethersulfone hollow fiber membranes seeded by primary human sinusoidal endothelial cells, stellate cells and hepatocytes in static and dynamic environment conditions [31].

The increasing use of 3D culture techniques in different research areas is accompanied by technical challenges for microscopy. Whereas 2D cultures can be conveniently analyzed by almost any kind of imaging, 3D culture systems have to be optimized and specifically prepared for permanent control of their growth and development.

Therefore, the detailed comparative study of the 2D vs 3D bioreactors for cell culture revealed that:

1) 2D monolayer bioreactors are useful for differentiation in the presence of growth factors and shear; gradient studies; and spatiotemporal pulses and steps.
2) 3D hydrogel bioreactors are useful for maintaining cells in undifferentiated state with high cell viability, and differentiation in the presence of growth factors.

3) 3D flow-perfuse bioreactors can be designed on different physical approaches for cell levitation (microgravity conditions) that will be discussed in the following sections.

4 Levitation of Cells in 3D Bioreactors

4.1 Spinner Flasks Bioreactors

Spinner flasks are simple bioreactor systems composed of a glass or plastic vessel in which cell/scaffold constructs are attached to vertical needles hanging from the top of the vessel and immersed in the culture medium. Different type and number of rotating elements can be used (Fig. 1a–d). The top of the vessel is usually used for gas exchange and medium oxygenation. Mixing of the medium is maintained with a stir bar at the bottom of the vessel or other mixing mechanisms. The convective forces generated during stirring mitigate the nutrient concentration gradients at the surface of the cell/scaffold constructs and produce turbulences that enhance mass transport toward the center of the samples. Spinner flasks were traditionally used to support large biomass growth and have recently been exploited for tissue engineering applications, including the culture and maturation of bone tissue substitutes using human osteo-competent cells derived from adult tissues.

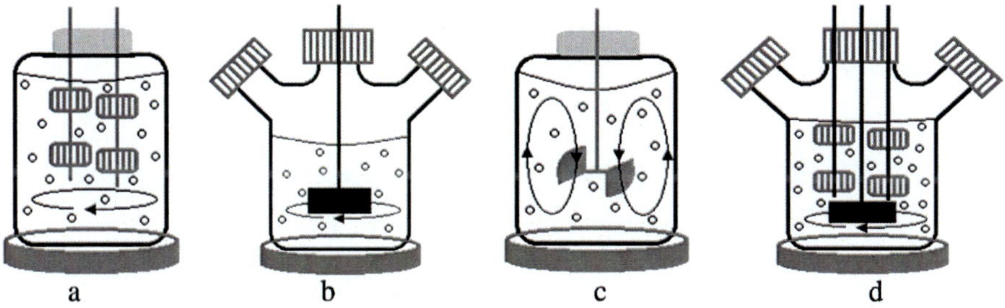

Fig. 1. Spinner flasks bioreactors with 1 (a) and 4 (b) scaffolds, 1 fan (c) and 4 scaffolds and fan (d) rotated.

4.2 Rotating Wall Vessel (RWV) Bioreactors

NASA initially developed the RWV bioreactor to protect cell cultures from the high shear forces generated during the launch and landing of the space shuttle. When the device was tested for cells in suspension on Earth, cells were seen to aggregate and form structures resembling tissues. These observations led to the possibility that the bioreactor might be used to study co-cultures of multiple cell types and the association of proliferation and differentiation during the early steps of tissue formation [32]. The RWV bioreactors have been used for modeling the microgravity environment encountered in space and to investigate growth, regulatory and structural processes in the microgravity-affected cultures [33, 34]. Its geometry and rotational speed was based on detailed **CFD computations** of the velocity fields in the vessel in order to get the uniform shear stress conditions for the cells [19, 35]. Later a spherical bioreactor with **two orthogonal axis of rotation** for better mixing has been designed and tested [36].

The RWV bioreactor was estimated as a useful tool to create a modeled microgravity, low fluid-shear environment that provides the necessary oxygenation and nutrients for development and polarization. In this environment, cells were observed to form cellular structures and features not readily expressed in conventional monolayer cell culture. Since its development, the RWV bioreactor has been utilized for the study of cellular and microbial gene expression in microgravity, cellular differentiation, host-pathogen interactions and tissue engineering [37–40]. The rotating wall vessel bioreactor was designed by NASA for experiments on cell culture in simulated microgravity conditions [33] (Fig. 2a, b). This bioreactor is com- posed of a horizontal cylindrical vessel rotating around a concentric tubular silicon membrane. The culture is located in the vessel while media oxygenation is provided via the membrane. The velocity of rotation equals to the sedimentation rate of the cells in the gravity field. Alternative configurations with a scaffold attached to the rotating wall have also been reported as the rotating bed bioreactor. In some designs the inner cylinder is also rotating (Fig. 2b). To date, a wide variety of cell lines and tissues has been successfully cultured and reproduced in the RWV bioreactor: placental, cervical, vaginal, neuronal and lung aggregates, as well as modeling virus-host interactions in various tissue settings [41]. Rotating wall vessels are limited to the cultures of small cell/scaffold constructs since they do not support optimal mass trans- port inside the construct core. In addition, due to the low range of values of shear stress imparted to the cells, these systems may not be efficient in promoting robust stress-driven differentiation [42].

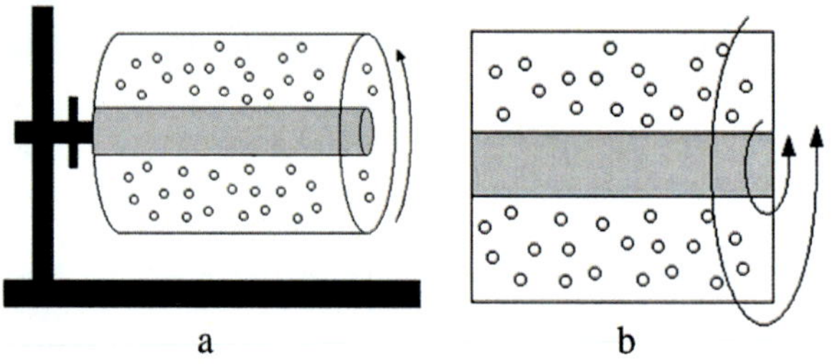

Fig. 2. Rotating wall bioreactors with outer (a), and outer and inner (b) rotation walls.

RWV bioreactors have been used to study the influence of microgravity the chon- drogenic differentiation of human adipose-derived MSCs (Fig. 3), which were cultured in pellets with or without the chondrogenic growth factor TGF-b1 [43]. MSCs are multi- potent, and their chondrogenesis is affected by mechanical stimulation. The histological results showed that microgravity had a synergistic effect on chondrogenesis with TGF- b1. Inhibition of p38 activity with SB203580 suppressed chondrocyte- specific gene expression and matrix production. These findings suggest that the p38 MAPK signal acts as an essential mediator in the microgravity induced chondrogene- sis of ADSCs.

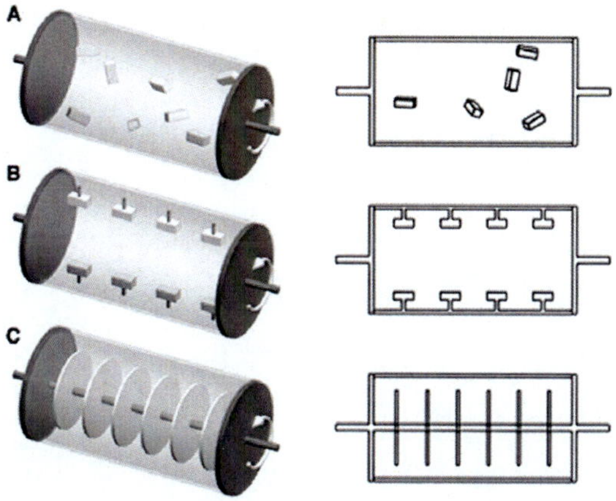

Fig. 3. Modifications of the RWV design: cultivation in a free-fall manner (a); scaffolds are attached to the outer vessel wall during cultivation (b); rotating bed bioreactor with cells seeded on discs rotating on the horizontal axis (c) (adapted from [43]).

4.3 Buoyant Levitation

It can be provided by using a feeding liquid of the same density as the cells have [44]. The neutral particles can be added to the liquid keeping the osmotic pressure and pH needed for the cells.

4.4 Perfusion Bioreactors (Fluid Dynamic Levitation)

The levitation can be achieved by floating the cells in the upward streams of liquid or gas. Perfusion bioreactors are culture systems composed of one or more perfusion chambers where the cell/scaffold constructs are placed, a medium reservoir, a tubing circuit and a pump enabling mass transport of nutrients and oxygen throughout the perfusion chamber (Fig. 4a,b). Perfusion bioreactors are broadly classified into indirect (Fig. 4a) or direct (Fig. 4b) systems, depending on whether the culture medium is perfused around or throughout the cell/scaffold constructs.

For a small number of cells the Bernoulli law based levitation can be used, but the influence of permanent rotation of the cell in the stream on the cell growth and development must be separately studied in experiments. Gas film levitation enables the levitation of an object against gravitational force by floating it on a thin gas film formed by gas flow through a porous membrane.

Indirect Perfusion Bioreactors. In the indirect perfusion bioreactors the cell/scaffold constructs are loosely placed in the perfusion chamber, and the culture medium preferentially follows (flow rates ~1–1.5 mL/min) the path of least resistance around the constructs (Fig. 4a), resulting in reduced mass transfer throughout the core of the samples.

Disadvantages: The convective forces generated by the perfusion pump mitigate the nutrient concentration gradients principally at the surface of the cell/scaffold constructs, thus limiting the number/size of the cells/aggregates that can be cultured.

Advantages: May represent valuable systems for the collective culture of a large number of small particulate cell/scaffold constructs [45, 46] that could be then assembled to repair large volumes of tissues.

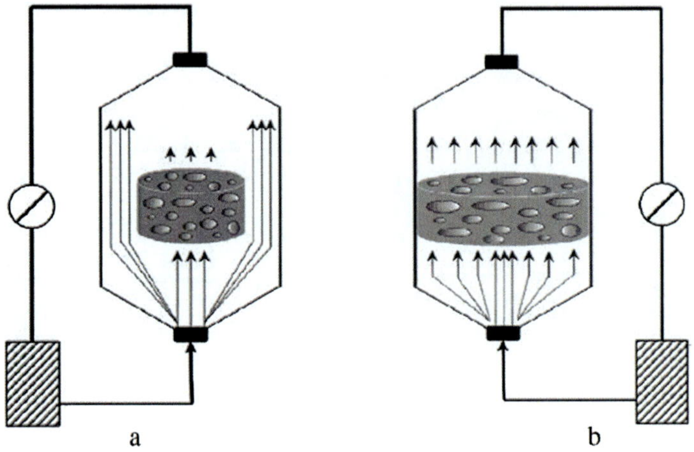

Fig. 4. Schematic representation of an indirect (a) and a direct (b) perfusion bioreactors with chambers (i), the cell/scaffold constructs (ii), the culture medium reservoirs (iii), the peristaltic pumps (iv) and the tubing systems (v).

Direct Perfusion Bioreactors. In direct perfusion bioreactors the cell/scaffold constructs are placed in the perfusion chamber in a press-fit fashion so that the culture medium is forced to pass through the center of the samples (Fig. 4b). This type of bioreactors has extensively been used to engineer bone substitutes using a combination of different human osteocompetent cells and biomaterial scaffolds.

Compression Bioreactors and Combined Systems. Compression bioreactors were intended to mimic the bone [32] and cartilage [47] physiological environment *in vitro*, characterized by repeated mechanical stimulation required for functional bone regeneration. These culture systems consist of a motor, a system providing linear motion and a compression chamber in which one or more pistons apply static or dynamic compressive loads directly to the cell/scaffold constructs (Fig. 5). In these systems the static or dynamic compressive loads are applied directly to the cells/constructs and controlled by a motor and a system that provides linear motion [42]. The combined units use perfusion + rotation or perfusion + compression systems for cell levitation, oxygenetion and nutrition.

4.5 Optic Levitation

Fixation of the cells can be produced in optical traps by light (radiation pressure) of a low-intensity laser [48–51]. Due to the high light reflection properties of many cells,

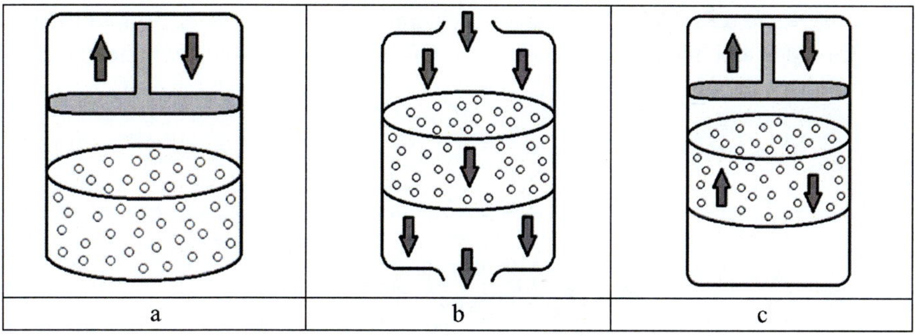

Fig. 5. Schematic representation of compression bioreactors with different types of compression chamber.

optical traps require very high light intensities of around 100 mW of laser light to trap and move cells. This high light intensity produces large local heat fluxes, which can be cytotoxic. The corresponding equipment is expensive.

4.6 Electromagnetic Levitation

Electric (EF) and electromagnetic (EMF) fields have been applied for bone regeneration purposes in patients with, for example, osteoporosis and non-unions as well as supportive therapy during limb lengthening and revision alloarthroplasty procedures for the last three decades [43, 52–56]. EMFs have been shown to significantly reduce the loss of bone mass and to accelerate bone formation in vivo [57]. Endogenous EMF also arise from muscle contractions [58]. The electric potentials generated by mechanical deformation in bone cause piezoelectricity. When bone is fractured, electrons migrate to the injured site, causing a negative potential. Vibrations of human muscles induce mechanical strains and currents of specific frequencies. Frequencies in the ranges of $f = 5–30$ Hz and $f < 10$ Hz were observed during postural muscle activity and walking, respectively [59]. Bone cells also exhibit a strong frequency selectivity with EMF effectiveness peaking at $f = 15$ Hz [60]. Experimental studies suggest that EMFs affect different subcellular proliferation- and differentiation-related signaling pathways, for example, those including parathyroid hormone and adenosine A2A receptor, resulting in conformation changes or in increase of the receptor density [61].

To utilize these effects for bone tissue engineering, EMF-based bioreactors (Fig. 6) were designed. Typically, these systems consist of Helmholtz coils powered by a PEMF generator. The cell/ scaffold construct is positioned between two Helmholtz coils and an EMF of a defined intensity is applied. In vitro studies showed that EMFs induce and enhance osteogenesis in human MSCs [62, 63] and osteoblasts [64–66]. A simple EMF-based bioreactor system with a standard well plate and two parallel Helmholtz coils being kept in a PMMA tube has also been developed [67]. The aplied PEMF frequency used in that study was 75 Hz with an intensity of 2 mT and the magnetic field was measured using a Hall Effect transverse gaussmeter probe. PEMF- stimulated human sarcoma osteogenic-2 cells exhibited increased mineralization and gene expression of decorin, OC, OPN, TGF-b, and Col1. In a study by Schwartz and co-workers, human

MSCs cultivated on calcium phosphate discs demonstrated de- creased proliferation but enhanced ALP activity and protein levels of OC and TGF-b in the combined presence of BMP-2 and PEMF stimulation [62].

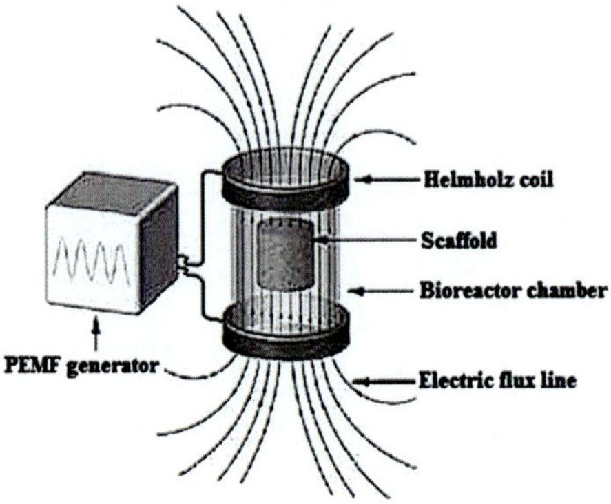

Fig. 6. Schematic illustration of a bioreactor system based on pulsed electromagnetic field.

It was shown, superior mineralization and expression of BMP-2 and BMP-4 genes can be achieved with PEMF stimulation in rat osteoblasts [64]. The device utilized a saw tooth waveform consisting of 4.5-ms bursts of pulses, repeating at a rate of 15 Hz. Increased proliferation and alkaline phosphatase activity were reported by Tsai et al., who stimulated rat osteoblasts seeded on poly(lactic-co-glycolic acid) scaffolds with PEMFs at a frequency of 7.5 Hz [68].

In that way, the use of EMF-based bioreactor systems for bone tissue engineering resulted in enhanced osteogenic differentiation of cell/scaffold constructs compared to static cultivation. Interestingly, as observed in some studies, PEMF also stimulated proliferation of osteoprogenitor cells. The high initial equipment costs required for PEMF-based bioreactor systems denote a major disadvantage. On the other hand, the noninvasiveness of PEMF-based systems is clearly advantageous with respect to handling and potential good manufacturing practice approval.

The electrostatic levitation can be supported by a non-uniform electric field to counteract gravitational force by the electrostatic ponderomotive force (Fig. 6). The strength E of the electric field is computed on known density and electric charge of the cells, viscosity of fluids and other electromechanical parameters to obtain the cell levitation condition between the electric, gravitation, buoyancy, Stokes, multiphase and other forces.

4.7 Dielectromagnetic Levitation

The subjects with large amount of water or other diamagnetic media can levitate in the strong magnetic field because diamagnets repel, and are repelled by a strong magnetic field. In the famous experiments on diamagnetic levitation of small frogs and mice the magnetic field $B = 13–17$ T have been used. The strong magnetic field of superconducting electromagnets is needed for macrovolumes of the cell cultures, while for the microfluidic purposes the magnetic films of NdFeB and SmCo can be used [69]. There are a number of commercially produced room temperature bore superconducting magnets designed for long-term magnetic levitation of diamagnetic samples. Size, geometry and force distribution needed for cell levitation and proper fluid mixing are based on CFD computations. Diamagnetic levitation can form quite dense populations of cells and their assembling into the microtissues. Additional aerodynamic stirring may be needed.

Diamagnetic levitation (B = 12.3 T) of bacterial cells during 18 h revealed absence of sedimentation and convective stirring of the basic fluid by the magnetic force [70]. It was demonstrated the diamagnetic levitation increased the rate of cell growth. Microarray gene analysis shows that the increase in growth rate is owing to enhanced oxygen availability, probably by enhancement of oxygen availability due to convection stirring oxygen around the liquid culture. Diamagnetic levitation of osteoblasts (B = 12.5 and 17 T) stimulated gene up-regulation due to intracellular stress or strain due to magnetic striction of the cells [71]. Magnetically levitated human cells showed similar protein expression profiles being located in a hydrogel consisting of gold, magnetic iron oxide nanoparticles and filamentous bacteriophage [72]. Magnetic field B = 7.6 T of a diamagnetic trap occurred selective stimulus specific effects on cell growth, cell cycle, and gene expression of yeast cells [73].

Exposition of the fruit fly D. melanogaster in the 0 g physical conditions generated by B = 16.5 T magnetic field during 22 days produced a delay in the development of the fruit flies from embryo to adult [74]. Microarray analysis indicated changes in overall gene expression after the exposition.

4.8 Magnetic Levitation

Weak magnetic forces generated by permanent magnet or electromagnet can be used for levitation of cells with paramagnetic properties (weak positive magnetic moment) or cells with magnetic nanoparticles attached to their surfaces.

Cells in monolayer culture are incubated in culture medium with nanoshuttle containing magnetic nanoparticles [75, 76]. After incubation with nanoshuttle, cells are detached and transferred to petri dishes and magnet(s) is(are) attached on top of the culture plate (Fig. 7). Within minutes, cells rise to the air–liquid interface being still suspended in the liquid; then within 24 hours, they self-assemble into 3D con- figurations. Magnetic force is chosen depending on the cell size, density and physical properties. Magnetic nanoparticle levitation systems with different magnetic forces are designed to fit 6-, 24- and 96-well tissue culture plates (Fig.7). The system was used for human glioblastoma cells grown for 48 hours in 1 mm aggregate. With magnetic drive in place, the aggregate levitates at the air–liquid interface. Following removal of magnetic drive, the 3D aggregate settles on the bottom of the Petri dish. The time of sedimentation can be computed easily.

4.9 Comparison of Different Levitation Techniques

Comparison of different above discussed levitation techniques is given in the Table 1.

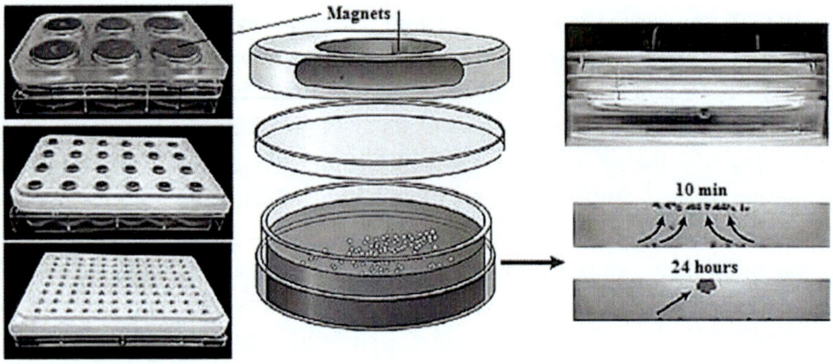

Fig. 7. Magnetic nanoparticles levitation bioreactors (adapted from [75]).

Table 1. Table Comparison of different existing cell levitation techniques.

Applied force	Equipment	3D resolution	Effect	Number of cells	Volume
Optic	Laser + microscopy	High	Low heating	Small	pL
Electric	Microelectrodes	Good	Heating	Single	mcL-mL
Magnetic	Magnets	Moderate	Moderate	Single	mcL-mL
Ultrasound	Resonance chamber	Low	Negligible	Large	mL

4.10 Influence of Microgravity/Levitation on Cell Physiology

How microgravity affects the biology of human cells and the formation of 3D cell cultures in real and simulated microgravity (r- and s-μg) is currently a hot topic in biomedicine [77].

Cytoskeletal Changes in Microgravity. As it was found in numerous orbital experiments, nearly all cell types exposed to microgravity is the alteration of cytoskeletal elements: actin, microfilaments and microtubules [78]. Disorganization of basic cellular architecture can affect activities ranging from cell signaling and migration to cell cycling and apoptosis; both gravity and spatial geometry affect the self-organization of microtubule networks [79, 80]. It was shown; tubulin assembles into microtubule structures in a gravity-independent manner. However, the organization of microtubule networks occurs by gravity-dependent reaction–diffusion processes, which are characterized by dynamic growth and the shortening of microtubules; conditions of reduced gravity drive this process to produce alternative orientations relative to those that occur in 1 g. Sample vessel geometry also influenced microtubule self-organization. Space experiments conducted in oval (egg-shaped) containers resulted in reduced microtubule organization compared with 1 g controls, whereas those conducted in rectangular containers exhibited almost no microtubule organization. Similar findings were observed using either clinorotation (i.e. clinostat) or magnetic levitation [81]. Cellular biophysical actions that affect tumor cell metastatic capacity, including migration, adhesion and invasion, are influenced by cytoskeletal organization [82–84]. Because microtubules provide architectural support for cellular organization, shape, motility and replication, it is tempting to speculate that disorder caused by exposure to microgravity has the potential to substantially affect cell growth and function.

5 Optimization of Bioreactors and Protocols: Trial-And-Error vs Computational Fluid Dynamics Approaches

Traditionally bioreactor designs and operation regimes were chosen by trial-and-error approaches, which are time consuming and often result in suboptimal performance and poor reproducibility [85]. To overcome such limitations and allow future translation of bone-engineered products, over the last years mathematical models and computer simulation techniques have been applied to an increasing extent to bioreactor systems in order to determine a priori the most favorable parameters for functional tissue regeneration from a set of available alternatives (see review in [19, 86]). Simulations in tissue engineering are usually divided into simulation of the biophysical environments (distribution of all physical forces) [87, 88] or biological environments (including dynamics of nutrient transport, tissue growth, matrix deposition and morphological evolution) [89], or both environments [90].

As opposite to cumbersome optical measuring techniques, such as for examples laser Doppler anemometry, particle image velocimetry and planar laser-induced fluorescence, computational fluid dynamics (CFD) softwares are now extensively used to provide simulations of the hydrodynamic environment of bioreactor systems and the factors influencing it. Distribution of mechanical forces, nutrient consumption and kinetics of tissue growth highly depend on mass transport, scaffold architecture and properties, as well as bioreactor design and operation regimes. In this view, given specific models of scaffolds filled with biological material, which can be developed using finite element methods, CFD simulations can be used to give recommendations about key elements of bioreactor design including shape, inlet and outlet location and size, and operation regimes, with the aim of providing adequate hydrodynamic environments for optimal cell culture.

In the FPB the cells experience gravity, buoyancy, centripetal, electric, magnetic (if any), viscous shear stress, viscoelastic interactions with other particles and solid surfaces. As a result, the particles will move along complex trajectories. Due to cell proliferation and growing cell concentrations, the rheology of concentrated suspensions with interacting particles with its shear rate, concentration and temperature de- pendent viscosity will be dominating. Then even in the case of slow wall rotation/fluid perfusion at low Reynolds numbers, the fluctuating motion will become irregular and chaotic. Nevertheless the long history of the RWV technique development, the detailed CFD computations on multicomponent and multiphase systems have been made possible during the last decade only thanks to high performance, high-availability cluster computing technologies. Even in a simple RWV design (Fig. 2a, b), the fluid dynamics based computed trajectories corresponded to the experimental ones at a quite narrow set of material parameters (particle size, shape, density, viscoelasticity, electric charge; fluid density, viscosity, electric conductivity, dielectric permeability), geometry of the bioreactor and flow regimes [19, 86, 91]. The collective dynamics of the particles is quite sensitive for cell size and mechanical properties, and at some combinations of the parameters the mechanical forces pro- mote hydrodynamic rapprochement and interaction. When formation of spheroid cellular aggregates is preferable, one class of bioreactors is needed. When the cell formation and growth in a cloud-type structure is essential for this type of cells or transplantation purposes, different fluid mechanics demands must be formulated on the CFD-based computer-assisted approach.

In the case of compression bioreactors, the **dynamic loading of cell/scaffold** constructs results in cyclic deformation of the scaffold and cellular material, which in turn affects the hydrodynamic behavior of the system, the distribution and intensity of the physical forces throughout the constructs over time, and the cellular response. Therefore, rigorous models must be developed using different numerical methods to specifically account for these fluid-structure interactions [92], and better predict formation and differentiation of tissues under specific loading conditions, which could also help in the development of more appropriate scaffold and bioreactor designs. The application of CFD simulations therefore plays an important role in the design of perfusion bioreactor configurations to engineer bone substitutes of clinical relevance, therefore reducing time and cost of product development for therapeutic translation, especially in clinical situations where customized bioreactors must be developed for functional regeneration of anatomically-shaped bone substitutes.

6 Conclusions

Urgent needs of modern regenerative medicine need novel next generation technologies for cell culture and tissue engineering. The thorough analysis of the patents and papers published shown that a big variety of 2D and 3D cell culture systems in fluid volumes and scaffolds have been designed, tested and approved during the last decades; nevertheless the individual cell and purpose oriented design and bioreactors and treatment protocol remains of a great importance for medicine and biology. In the in vivo conditions cells communicate with ECM and other cells in 3D space through biochemical, mechanical and electric signals that determines tissue/organ homeostasis. The cells are sensitive to mechanical stresses, and their biomechanical properties got pathologically changed by improper mechanical environment, which the cells can adapt to and control. Therefore, cell culture and tissue engineering conditions must correspond to natural biochemical, biophysical and biomechanical environments of the cell chosen. Cell culture at 3D dynamic conditions is preferable than 2D static conditions due to a number of reasons and factors that have been presented in the paper.

Traditionally bioreactor designs and operation regimes were chosen by trial-and-error approaches, while recent problems need efficient quantitative computation techniques which are accessible now due thanks to high-performance, high availability cluster computing technologies. Due to cell proliferation and increased cell concentrations, the rheology of concentrated suspensions with interacting particles with its shear rate, concentration and temperature dependent viscosity must be taken into account.

The material parameters (particle size, shape, density, viscoelasticity, electric charge; fluid density, viscosity, electric conductivity, dielectric permeability), geometry of the bioreactor and flow regimes are the main determinants of the desirable cell motion and proliferation behavior;

Additional experimental data on mechanical, electrical and other properties of single cells and cell suspensions at different concentrations are needed for the design/protocol optimization purposes.

References

1. Atala, A. (ed.): Foundations of Regenerative Medicine. Clinical and Therapeutic Applications. Academic Press. New York (2011)
2. Bhattacharya, N., Stubblefield, P.G. (eds.): Regenerative Medicine Using Non-Fetal Sources of Stem Cells. Springer, London (2015)
3. Braddock, M., Houston, P., Campbell, C., Ashcroft, P.: Born again bone: tissue engineering for bone repair. News Physiol. Sci. **16**, 208–213 (2001)
4. Frohlich, M., Grayson, W.L., Wan, L.Q., et al.: Tissue engineered bone grafts: biological requirements, tissue culture and clinical relevance. Curr. Stem Cell Res. Therapy **3**, 254–264 (2008)
5. Marolt, D., Knezevic, M., Vunjak-Novakovic, G.: Bone tissue engineering with human stem cells. Stem Cell Res. Therapy **1**, 10–16 (2010)
6. Atkinsonm, K. (ed.): The Biology and Therapeutic Application of Mesenchymal Cells. John Wiley & Sons, Inc., Hoboken (2016)
7. Ashoka, A., Choudhury, D., Fang, Y., Hunzikerac, W.: Towards manufacturing of human organoids. Biotechnol. Adv. **39**(2), 107460 (2020)
8. Bissell, M.J., Radisky, D.C., Rizki, A., et al.: The organizing principle: microenvironmental influences in the normal and malignant breast. Differentiation **70**(9–10), 537–546 (2002)
9. Martin, I., Wendt, D., Heberer, M.: The role of bioreactors in tissue engineering. Trends Biotechnol. **22**, 80–86 (2004)
10. Bayir, E., Sahinler, M., Celtikoglu, M.M., Sendemir, A.: Bioreactors in tissue engineering: mimicking the microenvironment. In: Biomaterials for Organ and Tissue Regeneration New Technologies and Future Prospects. Woodhead Publishing. Series in Biomaterials, pp. 709–752 (2020)
11. Lim, K.-T., Patel, D.K., Seonwoo, H., et al.: A fully automated bioreactor system for precise control of stem cell proliferation and differentiation. Biochem. Eng. J. **150**, 107258 (2019)
12. Pedersen, J.A., Swartz, M.A.: Mechanobiology in the third dimension. Ann. Biomed. Eng. **33**, 1469–1490 (2005)
13. Wang, J.H., Thampatty, B.P.: An introductory review of cell mechanobiology. Biomech. Model. Mechanobiol. **5**, 1–16 (2006)
14. Rutkowski, J.M., Swartz, M.A.: A driving force for change: interstitial flow as a morphoregulator. Trends Cell Biol. **17**(1), 44–50 (2007)
15. Wendt, D., Marsano, A., Jakob, M., et al.: Oscillating perfusion of cell suspensions through three-dimensional scaffolds enhances cell seeding efficiency and uniformity. Biotechnol. Bioeng. **84**, 205–214 (2003)
16. Zhao, F., Ma, T.: Perfusion bioreactor system for human mesenchymal stem cell tissue engineering: dynamic cell seeding and construct development. Biotechnol. Bioeng. **91**, 482–493 (2005)
17. Huang, C., Ogawa, R.: Effect of hydrostatic pressure on bone regeneration using human mesenchymal stem cells. Tissue Eng. Part A **18**, 2106–2113 (2012)

18. Weinbaum, S., Cowin, S.C., Zeng, Y.: A model for the excitation of osteocytes by mechanical loading-induced bone fluid shear stresses. J. Biomech. **27**(3), 339–360 (1994)
19. Delafosse, A., Calvo, S., Collignon, M.-L., et al.: Euler-Lagrange approach to model heterogeneities in stirred tank bioreactors – comparison to experimental flow characterization and particle tracking. Chem. Engin. Sci. **134**(29), 457–466 (2015)
20. Egger, D., Fischer, M., Clementi, A., et al.: Development and characterization of a parallelizable perfusion bioreactor for 3D cell culture. Bioengineering **4**, 51 (2017)
21. Pessoa, D.R., Finkler, A.T.J., Machado, A.V.L., et al.: CFD simulation of a packed-bed solid-state fermentation bioreactor. Appl. Math. Model. **70**, 439–458 (2019)
22. Loubière, C., Delafosse, A., Guedon, E., et al.: Dimensional analysis and CFD simulations of microcarrier 'just-suspended' state in mesenchymal stromal cells bioreactors. Chem. Engin. Sci. **203**, 464–474 (2019)
23. Keiger, D.: Moving cancer research out of the Petri dish and into the third dimension. Johns Hopkins Mag. **53** (2013)
24. Casser-Bette, M., Murray, A.B., Closs, E.I., et al.: Bone formation by osteoblast-like cells in a three-dimensional cell culture. Calcif. Tissue Int. **46**, 46–56 (1990)
25. Sourla, A., Doillon, C., Koutsilieris, M.: Three dimensional type I collagen gel system containing MG-63 osteoblast-like cells as a model for studying local bone reaction caused by metastatic cancer cells. Anticancer Res. **16**, 2773–2780 (1996)
26. Mizuno, M., Shindo, M., Kobayashi, D., et al.: Osteogenesis by bone stromal cells maintained on type I collagen matrix gels in vivo. Bone **20**, 101–107 (1997)
27. Hoffman, R.M.: 3D Sponge-matrix histoculture: an overview. Methods Mol. Biol. **1760**, 11–17 (2018)
28. Hendijani, F.: Explant culture: an advantageous method for isolation of mesenchymal stem cells from human tissues. Cell Prolif. **50**, e12334 (2017)
29. Haycock, J.W.: 3D cell culture: a review of current approaches and techniques. Methods Mol. Biol. **695**, 1–15 (2011)
30. Li, X.J., Valadez, A.V., Zuo, P., Nie, Z.: Microfluidic 3D cell culture: potential application for tissue-based bioassays. Bioanalysis **4**, 1509–1525 (2012)
31. Ahmed, H.M.M., Salerno, S., Morelli, S., et al.: 3D liver membrane system by co-culturing human hepatocytes, sinusoidal endothelial and stellate cells. Biofabrication **9**, 025022 (2017)
32. Granet, C., Laroche, N.L. Vico, C. Alexandre, M.H.: Lafage-proust rotating-wall vessels, promising bioreactors for osteoblastic cell culture: comparison with other 3D conditions. Med. Biol. Eng. Comput. **36**(4), 513–519 (1998)
33. Schwarz, R.P., Goodwin, T.J., Wolf, D.A.: Cell culture for three-dimensional modeling in rotating-wall vessels: an application of simulated microgravity. J. Tissue Cult. Methods **14**, 51–57 (1992)
34. Goodwin, T.J., Schroeder, W.F., Wolf, D.A., Moyer, M.P.: Rotating-wall vessel coculture of small intestine as a prelude to tissue modeling: aspects of simulated microgravity. Proc. Soc. Exp. Biol. Med. **202**, 181–192 (1993)
35. Tsao, Y.D., Goodwin, T.J., Wolf, D.A., Spaulding, G.F.: Responses of gravity level variations on the NASA/JSC bioreactor system. Physiologist **35**(1), S49-50 (1992)
36. Zhang, Z.-Y., Teoh, S.H., Chong, W.-S.: A biaxial rotating bioreactor for the culture of fetal mesenchymal stem cells for bone tissue engineering. Biomaterials **30**, 2694–2704 (2009)
37. Navran, S.: The application of low shear modeled microgravity to 3D cell biology and tissue engineering. Biotechnol. Annu. Rev. **14**, 275–296 (2008)
38. Barrila, J., Radtke, A.L., Crabbe, A., et al.: Organotypic 3d cell culture models: using the rotating wall vessel to study host-pathogen interactions. Nat. Rev. Microbiol. **8**, 791–801 (2010)
39. Barzegari, A., Saei, A.A.: An update to space biomedical research: tissue engineering in microgravity bioreactors. BioImpacts **2**, 23–32 (2012)

40. Grimm, D., Wehland, M., Pietsch, J., et al.: Growing tissues in real and simulated microgravity: new methods for tissue engineering. Tissue Eng. **20**, 555–566 (2014)
41. Gardner, J.K., Herbst-Kralovetz, M.H.: Three-dimensional rotating wall vessel-derived cell culture models for studying virus-host interactions. Viruses **8**, 304 (2016)
42. Sladkova, M., de Peppo, G.M.: Bioreactor systems for human bone tissue engineering. Processes **2**, 494–525 (2014)
43. Rauh, J., Milan, F., Gunther, K.-P., Stiehler, M.: Bioreactor systems for bone tissue engineering. Tissue Eng. Part B **17**(4), 263–280 (2011)
44. Anil-Inevi, M., Yaman, S., Yildiz, A.A., et al.: Biofabrication of in situ self assembled 3D cell cultures in a weightlessness environment generated using magnetic levitation. Sci. Rep. **8**, 7239 (2018)
45. de Peppo, G.M., Sladkova, M., Sjövall, P., et al.: Human embryonic stem cell-derived mesodermal progenitors display substantially increased tissue formation compared to human mesenchymal stem cells under dynamic culture conditions in a packed bed/column bioreactor. Tissue Eng. Part A **19**, 175–187 (2013)
46. David, B., Bonnefont-Rousselot, D., Oudina, K., et al.: Perfusion bioreactor for engineering bone constructs: an in vitro and in vivo study. Tissue Eng. **17**, 505–516 (2011)
47. Freed, L.E., Vunjak-Nopvakovic, G., Langer, R.: Cultivations of cell-polymer cartilage implants in bioreactors. J. Cell. Biochem. **51**, 257–264 (1993)
48. Grier, D.G.: A revolution in optical manipulation. Nature **424**(6950), 810–816 (2003)
49. Molloy, J.E.: Optical chopsticks: digital synthesis of multiple optical traps. Methods Cell Biol. **55**, 205–216 (2018)
50. Rodrigo, P.J., Daria V.R., Gluckstad, J.: Real-time three-dimensional optical micromanipulation of multiple particles and living cells. Opt. Lett. **29**(19), 2270–2272 (2019)
51. Winkleman, A., Gudiksen, K.L., Ryan, D., et al.: A magnetic trap for living cells suspended in a paramagnetic buffer. Appl. Phys. Lett. **85**(12), 2411–2413 (2004)
52. Bassett, C.A., Pilla, A.A., Pawluk, R.J.: A non-operative salvage of surgically-resistant pseudarthroses and nonunions by pulsing electromagnetic fields. A preliminary report. Clin. Orthop. Relat. Res. **124** 128–133 (1977)
53. Gossling, H.R., Bernstein, R.A., Abbott, J.: Treatment of ununited tibial fractures: a comparison of surgery and pulsed electromagnetic fields. Orthopedics **15**, 711 (1992)
54. Eyres, K.S., Saleh, M., Kanis, J.A.: Effect of pulsed electromagnetic fields on bone formation and bone loss during limb lengthening. Bone **18**, 505 (1996)
55. Aaron, R.K., Ciombor, D.M., Simon, B.J.: Treatment of nonunions with electric and electromagnetic fields. Clin. Orthop. Relat. Res. **419**, 21 (2004)
56. Dallari, D.: Effects of pulsed electromagnetic stimulation on patients undergoing hip revision prostheses: a randomized prospective double-blind study. Bioelectromag. **30**, 423 (2009)
57. Midura, R.J.: Pulsed electromagnetic field treatments enhance the healing of fibular osteotomies. J. Orthop. Res. **23**, 1035 (2005)
58. Funk, R.H., Monsees, T., Ozkucur, N.: Electromagnetic effectsFrom cell biology to medicine. Prog. Histochem. Cytochem. **43**, 177 (2009)
59. Antonsson, E.K., Mann, R.W.: The frequency content of gait. J. Biomech. **18**, 39 (1985)
60. Rubin, C.T., Donahue, H.J., Rubin, J.E., McLeod, K.J.: Optimization of electric field parameters for the control of bone remodeling: exploitation of an indigenous mechanism for the prevention of osteopenia. J. Bone Miner. Res. **8**(2), S573 (1993)
61. Aaron, R.K., Ciombor, D.M., Wang, S., Simon, B.: Clinical biophysics: the promotion of skeletal repair by physical forces. Ann. NY Acad. Sci. **1068**, 513–519 (2006)
62. Schwartz, Z.: Pulsed electromagnetic fields enhance BMP-2 dependent osteoblastic differentiation of human mesenchymal stem cells. J. Orthop. Res. **26**, 1250 (2008)

63. Sun, L.Y., Hsieh, D.K., Lin, P.C., et al.: Pulsed electromagnetic fields accelerate proliferation and osteogenic gene expression in human bone marrow mesenchymal stem cells during osteogenic differentiation. Bioelectromag. **31**, 209–218 (2010)
64. Bodamyali, T., Bhatt, B., Hughes, F.J., et al.: Pulsed electromagnetic fields simultaneously induce osteogenesis and upregulate trantranscription of bone morphogenetic proteins 2 and 4 in rat osteoblasts in vitro. Biochem. Biophys. Res. Commun. **250**, 458–464 (1998)
65. Wiesmann, H., Hartig, M., Stratmann, U., et al.: Electrical stimulation influences mineral formation of osteoblast-like cells in vitro. Biochim. Biophys. Acta. **1538**, 28–37 (2001)
66. Icaro, C.A., Casasco, M., Riva, F., et al.: Stimulation of osteoblast growth by an electromagnetic field in a model of bone-like construct. Eur. J. Histochem. **50**, 199–209 (2006)
67. Fassina, L.: Effects of electromagnetic stimulation on calcified matrix production by SAOS-2 cells over a polyurethane porous scaffold. Tissue Eng. **12**, 1985–1994 (2006)
68. Tsai, M.T., Chang, W.H., Chang, K., et al.: Pulsed electromagnetic fields affect osteoblast proliferation and differentiation in bone tissue engineering. Bioelectromagn. **28**, 519–526 (2017)
69. Walter, A., Marcoux, C., Desloges, B., et al.: Micro-patterning of NdFeB and SmCo magnet films for integration into micro-electro-mechanical-systems. J. Magn. Magnet. Mater. **321**, 590–594 (2009)
70. Dijkstra, C.E., Larkin, O.J., Anthony, P., et al.: Diamagnetic levitation enhances growth of liquid bacterial cultures by increasing oxygen availability. J. R. Soc. Interface **8**, 334–344 (2011)
71. Hammer, B.E., Kidder, L.S., Williams, P.C., Xu, W.W.: Magnetic levitation of MC3T3 osteoblast cells as a ground-based simulation of microgravity. Microgravity Sci. Technol. **21**(4), 311–318 (2009)
72. Souza, G.R., Molina, J.R., Raphael, R.M., et al.: Three-dimensional tissue culture based on magnetic cell levitation. Nat. Nanotechnol. **5**, 291–296 (2010)
73. Coleman, C., Gonzalez-Villalobos, R.A., Allen, P.L., et al.: Hammond diamagnetic levitation changes growth, cell cycle, and gene expression of *Saccharomyces cerevisiae*. Biotechnol. Bioengin. **98**(4), 854–863 (2007)
74. Herranz, R., Larkin, O.J., Dijkstra, C.E., et al.: Microgravity simulation by diamagnetic levitation: effects of a strong gradient magnetic field on the transcriptional profile of *Drosophila melanogaster*. BMC Genom. **13**, 52–59 (2021)
75. Becker, J.L., Souza, G.R.: Using space-based investigations to inform cancer research on Earth. Nat. Rev. Cancer **13**(5), 315–327 (2018)
76. Markx, G.H.: The use of electric fields in tissue engineering. Organogenesis **4**(1), 11–17 (2018)
77. Ulbrich, C., Wehland, M., Pietsch, J., et al.: The impact of simulated and real microgravity on bone cells and mesenchymal stem cells. Biomed. Res. Int. **2014**, 928507 (2014)
78. Lewis, M.L.: The cytoskeleton in spaceflown cells: an overview. Grav. Space Biol. Bull. **17**, 1–11 (2004)
79. Tabony, J.: Gravity dependence of microtubule selforganization. Grav. Space Biol. Bull. **17**, 13–25 (2004)
80. Papaseit, C., Pochon, N., Tabony, J.: Microtubule self-organization is gravity-dependent. Proc. Natl Acad. Sci. USA **97**, 8364–8368 (2000)
81. Glade, N., Beaugnon, E., Tabony, J.: Ground-based methods reproduce space-flight experiments and show that weak vibrations trigger microtubule self-organization. Biophys. Biochem. **121**, 1–6 (2006)
82. Korb, T.: Integrity of actin fibers and microtubules influences metastatic tumor adhesion. Exp. Cell Res. **299**, 236–247 (2004)

83. Suresh, S.: Biomechanics and biophysics of cancer cells. Acta Biomater. **3**, 413–438 (2007)
84. Kim, Y., Stolarska, M.A., Othmer, H.G.: The role of microenvironment in tumor growth and invasion. Prog. Biophys. Mol. Biol. **106**, 353–379 (2011)
85. Hutmacher, D.W., Singh, H.: Computational fluid dynamics for improved bioreactor design and 3D culture. Trends Biotechnol. **26**, 166–172 (2008)
86. Vetsch, J.R., Müller, R., Hofmann, S.: The evolution of simulation techniques for dynamic bone tissue engineering in bioreactors. J. Tissue Eng. Regen. Med. **9**(8), 903–917 (2015)
87. Sucosky, P., Osorio, D.F., Brown, J.B., Neitzel, G.P.: Fluid mechanics of a spinner-flask bioreactor. Biotechnol. Bioeng. **85**, 34–46 (2004)
88. Milan, J.L., Planell, J.A., Lacroix, D.: Simulation of bone tissue formation within a porous scaffold under dynamic compression. Biomech. Model. Mech. **9**, 583–596 (2010)
89. Galbusera, F., Cioffi, M., Raimondi, M.T., Pietrabissa, R., et al.: Computational modeling of combined cell population dynamics and oxygen transport in engineered tissue subject to interstitial perfusion. Comput. Methods Biomech. Biomed. Eng. **10**, 279–287 (2007)
90. Lappa, M.: Organic tissues in rotating bioreactors: Fluid-mechanical aspects, dynamic growth models, and morphological evolution. Biotechnol. Bioeng. **84**, 518–532 (2003)
91. Liu, T., Li, X., Sun, X., et al.: Analysis on forces and movement of cultivated particles in a rotating wall vessel bioreactor. Biochem. Engin. J. **18**(2), 97–104 (2004)
92. Verbruggen, S.W., Vaughan, T.J., McNamara, L.M.: Fluid flow in the osteocyte mechanical environment: a fluid-structure interaction approach. Biomech. Model. Mechanobiol. **13**, 85–97 (2014)

Assessing the Feasibility of Using Spherical Contact Pairs to Model the Contact Regions in the Joints of the Index Finger

Karol Kluza and Adam Ciszkiewicz(✉)

Faculty of Mechanical Engineering, Cracow University of Technology, Cracow, Poland
adam.ciszkiewicz@pk.edu.pl

Abstract. The index finger is a complex structure in the human body, which contains multiple joints. It is used for precise interaction with the environment. In the available studies, the joints of the finger are usually replaced with simple constraints, such as revolute or spherical joints. The main aim of this research was to assess whether the contact areas in the three joints of the finger could be replaced with simple spherical contact pairs. This was motivated by the fact that such contact pairs would allow for more accurate contact analysis than the simplified constraints, while still being computationally inexpensive. The research was performed using two computer tomography datasets, which were transformed into surface meshes of the bones. Parts of the bones, which corresponded to the contact areas in the joints, were selected and imported into Python. Then, sphere fitting was performed in a custom script, which also allowed for visualization of the obtained contact pairs – obtained from the fitted spheres. The pairs were then analyzed. The results showed that it might be possible to replace the contact areas in the joints of the index finger with simple spherical contact pairs. Moreover, in two of the joints, better results were obtained when using two contact pairs – medial and lateral – instead of one for the whole area.

Keywords: Multibody system method · Bone contact · Digit · Shape fitting

1 Introduction

The index finger is a complex multijoint structure in the human body, which is used for precise interaction with the environment. At the same time, it is also the most injured of the digits in the human hand [1].

The finger can be divided into three major joints: distal interphalangeal (DIP), proximal interphalangeal (PIP), and metacarpophalangeal (MCP). Each one of these joints connects two bones through cartilage and a complicated system of ligaments. The muscles and tendons additionally stabilize and articulate the joints. In most of the available studies, the index finger is usually treated as a structure with four degrees of freedom (DOF) and modeled under a multibody system framework [2–5]. Two of the DOFs are realized through simple hinge joints, which replace DIP and PIP. MCP is substituted with an universal joint of two DOFs. These simplified representations of the joints are based

A. Hadamus et al. (Eds.): BIOMECHANICS 2021, LNNS 328, pp. 100–109, 2022.
https://doi.org/10.1007/978-3-030-86297-8_9

on the analysis of the axes of rotation of the index finger. The approach works well in complex models, featuring multiple joints [6–8], as it is numerically efficient. Nevertheless, due to the simplifications, most of the individual elements of the finger joints, such as contact areas and ligaments, are lost in the modeling process, making these models insufficient for more detailedstudies regarding, for instance, stress distribution in the articular surfaces. In such cases, finite element approaches are usually applied [9–13]. It is not uncommon to see finite element models connected toa multibody framework, creating a hybrid approach [9].

While the finite element framework works well with complex and detailed models of human body joints, it is also very intensive numerically and numerical issues may occur when solving the models. On the other hand, a different approach to soft tissue modeling has been presented in [14–18] for the knee,hip,and ankle. In these studies, the cartilage is modeled under the multibody framework with simplified geometrical shapes, which create deformable contact pairs under the Hertzian contact formulation. This approach offers a more realistic contact model than the simplified constraints used to typically model the finger joint with the multibody system method. At the same time, it is also less computationally expensive than the finite element approach. To the best of our knowledge, such a method for contact description has not been applied to index finger joints.

Our aim in this research was to analyze whether the cartilage in the joints of the index finger could be substituted with simple spherical contact pairs, taking into account the anatomical features of the joints' bones. The geometrical analysis was performed using two three-dimensional models of the index finger and two sphere fitting procedures implemented in a *Python* script. The obtained contact pairs were rated based on the quality of the sphere fitting and the viability of the obtained contact pairs in an anatomical sense. Specific details regarding the procedure and the obtained results were included in the following sections.

2 Material and Methodology

The study was divided into several steps. In the first one, two three-dimensional surface models of the bones in the finger were obtained based on computer tomography (CT) scans. Then, parts of these meshes were selected and imported into a *Python* script, which was used to fit the spheres to them. For DIP and PIP, two contact configurations were considered – one contact pair for the whole cartilage and two contact pairs, medial and lateral, for the cartilage. In case of MCP, one contact pair was employed. The spheres obtained for each joint in the finger were analyzed in terms of the fit quality and of whether these spheres could form a proper contact pair to model the cartilage.

Note that the sphere fitting was performed on meshes, which represented the bone profiles of the joints. This was a simplification. In the actual joints, the bones interact through cartilage. In our case, the CT scans did not contain soft tissue, therefore we decided to use the bone profiles to obtain the contact pairs for the joints – for our purpose, this was acceptable. Similar approach was used before for other body joints [19, 20]. It is worth mentioning that thickness of the cartilage could be taken into account after obtaining the contact pairs, for instance, based on experimental data available in the literature [21–23].

The following sections of the paper contain a detailed description of the procedure used for obtaining and rating the contact pairs.

2.1 Obtaining Bone Profiles

As mentioned before, the bone profiles in this study were obtained based on two CT scans of the index finger. The scans were downloaded from *embodi3d.com*, where they were shared under the *Creative-Commons By Attribution* license. The details regarding the datasets were listed below:

- filename: Hand surface scan 1.0.0by Krishna Josyula at https://www.embodi3d.com/files/file/40118-hand-surface-scan/ under CC – Attribution license and accessed: 25.11.2020,
- filename: Kosci_dloni_3 1.0.0 by MABC at https://www.embodi3d.com/files/file/22155-kosci_dloni_3/ under CC – Attribution license and accessed: 25.11.2020.

The bone profiles for each dataset were obtained using an open-source computer program for processing medical images called *3D Slicer*. The software allowed for importing the CT scans, visualizing them in different projections, selecting a region of interest, and performing bone segmentation. The segmentation was initially carried out with thresholding, but the obtained images were then manually finetuned to remove any errors from thresholding. Finally, three-dimensional surface meshes of the bones were computed using the marching cubes algorithm. The two obtained models were presented in Fig. 1.

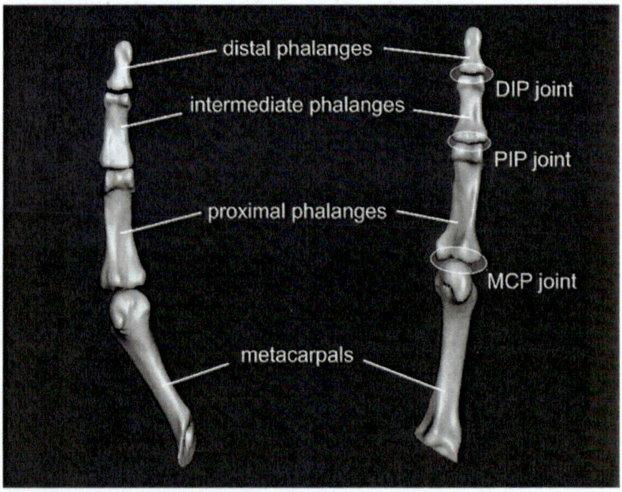

Fig. 1. The obtained three-dimensional surface meshes of the index finger in the two cases.

2.2 Extracting the Contact Regions for Contact Pair Fitting

The obtained meshes of the index finger were then further processed in *Meshmixer*, which allowed for selecting and cutting parts of the meshes. The software was used to select the regions in the bone, which represented the contact areas in the joints.

A sample of the selection process in *Meshmixer* was presented in Fig. 2.

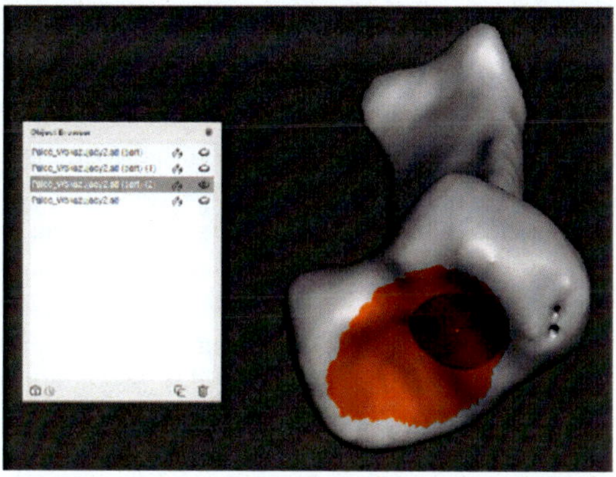

Fig. 2. Selecting parts of the bone mesh in *Meshmixer*.

The selection and cutting procedure was performed based on the analysis of the anatomy of the finger joints. Our main goal was to obtain the simplest possible contact pairs to substitute the contact areas in the joints. Therefore, we initially obtainedonly one spherical pair for each joint. Nevertheless, due to the complexity of the geometry of the DIP and the PIP, we decided to additionally subdivide their contact regions into two spherical contact pairs – lateral and medial. These two approaches were compared and analyzed in the results section.

The selected parts of the bone meshes – mesh cuts – were saved as STL files and loaded into a script in Python, which allowed for:

- sphere fitting for each mesh cut with optimization as presented in [24],
- visualization of the mesh cut along with the fitted spheres.

In total, 14 contact pairs were obtained using the aforementioned procedure (3 for DIP, 3 for PIP, and 1 for MCP – for each of the two 3D models). An example of an contact pair was shown in Fig. 3.

To evaluate the obtained contact pairs, we used two different criteria. Firstly, the visualized pairs were analyzed in terms of whether they formed valid contact pairs – their spheres did not intersect. In addition, we also computed parameters to judge the geometry and the quality of the sphere fit – such as the radius, the maximal deviation from the sphere, and the standard deviation.

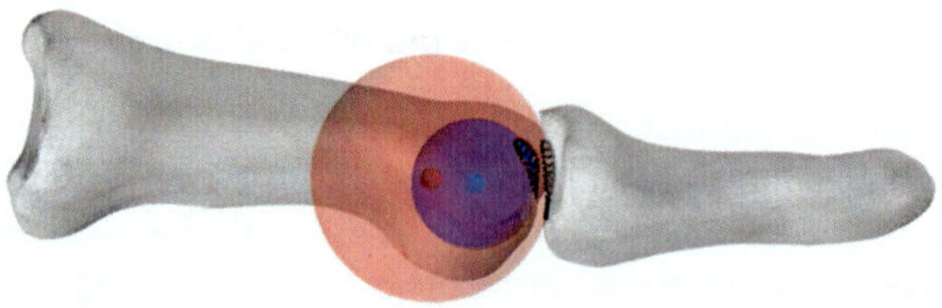

Fig. 3. One of the obtained contact pairs along with the meshes of the bones.

3 Results and Discussion

As mentioned in the method section, our goal was to use the simplest possible geometry to describe the contact regions in the joints of the index finger. Therefore, in our first trial, we employed a singular spherical contact pair for each joint – see the results presented in Figs. 4 and 5.

The approach worked well for the MCP, in which the bone shapes corresponded to spheres, as shown in Fig. 5. Nevertheless, this simplification proved to be incorrect for the DIP andPIP – see Fig. 4. In these cases the procedure returned mathematically correct pairs, but with very large dimensions – see Table 1, especially when compared to the assumed mesh cuts or the size of the joint itself. This meant that the subtle curvature of the bone profiles was completely lost – the procedure essentially flattened the whole surface to a large sphere, which was reflected by the relatively high value of the maximal difference between the points of the mesh cuts and the fitted spheres – see Table 1.This was true for both studied 3D models. Furthermore, in the second model, the ratio of radii in the contact pair for PIP was inversed.

To address the aforementioned issue, we subdivided the contact regions in DIP and PIP into two separate parts – medial, lateral – and used them to obtain two spherical pairs for these joints. This idea was partially inspired by similar applications for the knee and the ankle [15, 17].

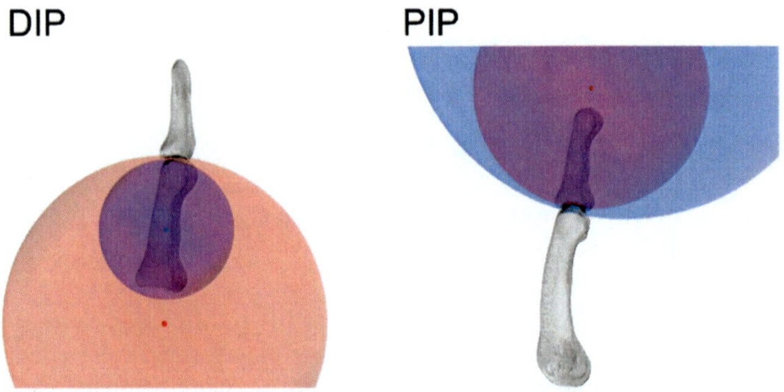

Fig. 4. The contact pairs along with their mesh cuts and meshes of the bones for the DIP and PIP joints in the index finger based on the second one of the considered models of the digit.

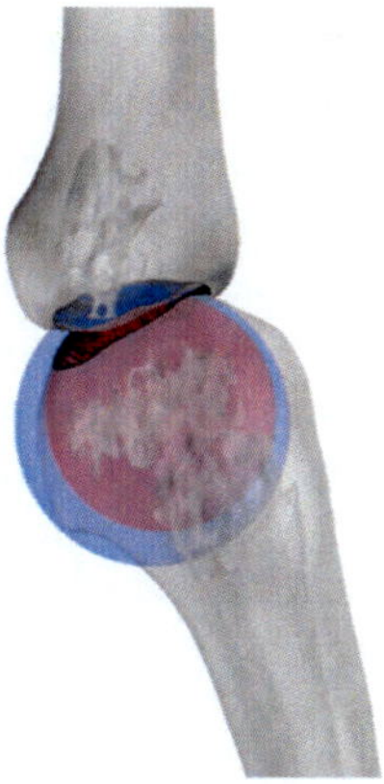

Fig. 5. The obtained contact pair for the MCP in the second model.

Table 1. The details regarding the spheres fitted to the three joints in the two models, where: ***J*** – the joint, ***M*** – the 3D model, ***id*** – the id of the contact pair, ***r*** – the radius of a sphere, ***std*** – the standard deviation, ***mx*** – the maximal difference between the points of the mesh cuts and the fitted sphere. Valid contact pairs, in which the upper radius was larger than the lower radius, were underlined in the ***id*** and ***r*** columns.

J	*M*	*id*	*r* [mm]	*std* [mm]	*mx* [mm]
DIP	1	*#1*	48.65	0.15	0.57
			29.76	0.27	1.22
	2	*#2*	30.45	0.16	0.36
			12.74	0.31	0.74
PIP	1	*#3*	57.31	0.36	1.01
			23.45	0.32	0.90
	2	*#4*	28.62	0.31	0.60
			54.44	0.41	1.04
MCP	1	*#5*	12.83	0.27	1.13
			9.57	0.13	0.52
	2	*#6*	9.23	0.19	0.59
			7.76	0.17	0.55

The dual-contact approach returned significantly better results, as seen in the standard deviation and maximal difference – see Table 2. Moreover, to further validate the conformity of this approach, we computed relative indicators comparing the radius, standard deviation and maximal difference between the single and dual contact approach – see Table 2. It is worth mentioning that in all cases these parameters were improved and in some cases the reduction percentage of the standard deviation and maximal difference could be as high as 88%. Furthermore, the radius of the spheres was reduced 5.36 times on average – the spheres were smaller and better fit to the curvature of the contact regions of the joints.

In terms of the viability of the obtained contact pairs, the visual assessment results were as follows. For the second model, all contact pairs obtained with the dual-contact approach formed anatomically proper ball-and-socket joints. Nonetheless, for the first model, two of the considered contact pairs were incorrect. Namely, the ratio of the radii between their spheres was reversed – the anatomically larger sphere (the socket) was smaller after fitting. We suspect that the reason for this issue was in the quality of the first scan or the subjective aspect of contact region selection on the bone meshes. This problem requires further study.

Table 2. Details regarding the spheres fitted to the DIP and PIP when their contact areas were separated into two contact pairs, where: ***J*** – the joint, ***M*** – the 3D model, ***id*** – the id of the contact pair, ***S*** – the side of the contact region, ***r*** – the radius of a sphere, ***std*** – the standard deviation, ***mx*** – the maximal difference between the points of the mesh cuts and the fitted sphere. The columns with "%" represent the ratio between the parameters (***r***, ***std***, ***mx***) obtained from a dual contact approach and the single contact approach (Table 1) – the lower the value, the better. Valid contact pairs, in which the upper radius was larger than the lower radius, were underlined in the ***id*** and ***r*** columns.

J	*M*	*S*	*id*	*r* [mm]	$r_{\%}$[%]	*std* [mm]	$std_{\%}$ [%]	*mx* [mm]	$mx_{\%}$ [%]
DIP	1	*A*	*#1A*	4.99	10.26	0.10	66.67	0.34	59.65
				5.98	20.09	0.13	48.15	0.52	42.62
		B	*#1B*	7.26	14.92	0.04	26.67	0.13	22.81
				4.46	14.99	0.06	22.22	0.14	11.48
	2	*A*	*#2A*	5.51	18.10	0.06	37.50	0.14	38.89
				2.89	22.68	0.14	45.16	0.38	51.35
		B	*#2B*	5.23	17.18	0.05	31.25	0.12	33.33
				2.74	21.51	0.04	12.90	0.10	13.51
PIP	1	*A*	*#3A*	7.95	13.87	0.14	38.89	0.46	45.54
				6.81	29.04	0.16	50.00	0.54	60.00
		B	*#3B*	7.62	13.30	0.16	44.44	0.35	34.65
				9.13	38.93	0.18	56.25	0.56	62.22
	2	*A*	*#4A*	4.51	15.76	0.06	19.35	0.16	26.67
				3.60	6.61	0.05	12.20	0.14	13.46
		B	*#4B*	8.91	31.13	0.08	25.81	0.19	31.67
				5.49	10.08	0.12	29.27	0.40	38.46

As mentioned above, the obtained contact pairs in the second model formed proper ball-and-socket joints, corresponding to the anatomy of the joints. These results for the DIP and PIP were presented in Fig. 6.

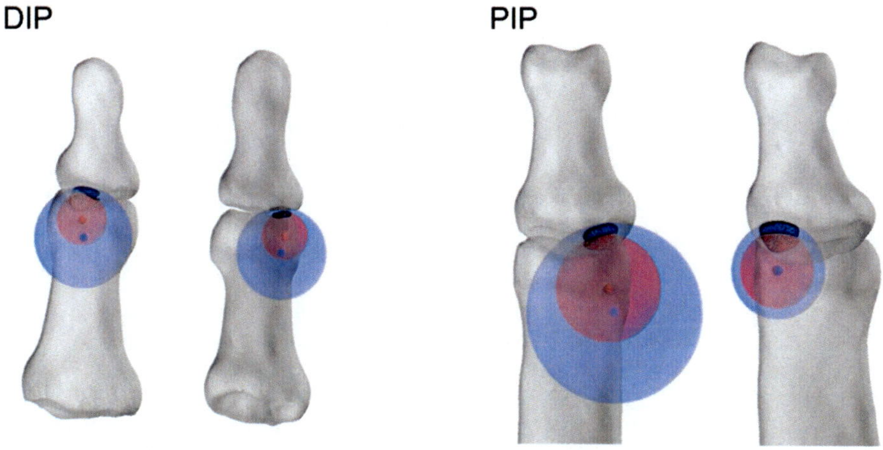

Fig. 6. The obtained contact pairs for DIP and PIP in the second model.

To summarize, the obtained results proved that it was possible to replace the contact region in MCP with a singular spherical contact pair. For the DIP and PIP, viable results were obtained when using two contact pairs – medial and lateral. The contact areas in these joints could not be simplified to a single contact pair. In total, almost 80% of the obtained pairs were viable.

4 Conclusion

The main aim of this study was the assessment of whether the contact areas in the joints of the index finger could be replaced with simple spherical contact pairs. The research was performed using two computer tomography datasets, which were transformed into surface meshes of the bones. Parts of the bones, which corresponded to the contact regions in the joints, were selected and imported into Python. Then, sphere fitting was performed in a custom script and the obtained contact pairs were visualized. The results showed that 11 out of the considered 14 contact pairs formed anatomically correct ball-and-socket joints. It was possible to replace the contact region in the MCP with only a singular contact pair. On the other hand, the DIP and PIP joints required two contact pairs – one on the medial and one on the lateral side. Unfortunately, the contact areas in these joints could not be simplified to a single contact pair.

References

1. White, W.L.: Why I hate the index finger. Orthop. Rev. **9**, 23–29 (1980). https://doi.org/10.1007/s11552-010-9285-5
2. Hu, D., Howard, D., Ren, L.: Biomechanical analysis of the human finger extensor mechanism during isometric pressing. PLoS One **9** (2014). https://doi.org/10.1371/journal.pone.0094533
3. Brook, N., Mizrahi, J., Shoham, M., Dayan, J.: A biomechanical model of index finger dynamics. Med. Eng. Phys. **17**, 54–63 (1995). https://doi.org/10.1016/1350-4533(95)90378-O
4. Qin, J., Lee, D., Li, Z., Chen, H., Dennerlein, J.T.: Estimating in vivo passive forces of the index finger muscles: Exploring model parameters. J. Biomech. **43**, 1358–1363 (2010). https://doi.org/10.1016/j.jbiomech.2010.01.014
5. Barry, A.J., Murray, W.M., Kamper, D.G.: Development of a dynamic index finger and thumb model to study impairment. J. Biomech. **77**, 206–210 (2018). https://doi.org/10.1016/j.jbiomech.2018.06.017
6. Laitenberger, M., Raison, M., Périé, D., Begon, M.: Refinement of the upper limb joint kinematics and dynamics using a subject-specific closed-loop forearm model. Multibody Sys.Dyn. **33**, 413–438 (2014). https://doi.org/10.1007/s11044-014-9421-z
7. Lemay, M.A., Crago, P.E.: A dynamic model for simulating movements of the elbow, forearm, and wrist. J. Biomech. **29**, 1319–1330 (1996). https://doi.org/10.1016/0021-9290(96)00026-7
8. Weinberg, A.M., Pietsch, I.T., Helm, M.B., Hesselbach, J., Tscherne, H.: A new kinematic model of pro- and supination of the human forearm. J. Biomech. **33**, 487–491 (2000). https://doi.org/10.1016/S0021-9290(99)00195-5
9. Faudot, B., Milan, J., De Monsabert, B.G., Le Corroller, T., Vigouroux, L.: Estimation of joint contact pressure in the index finger using a hybrid finite element musculoskeletal approach. Comput. Methods Biomech. Biomed. Eng. 1–11 (2020). https://doi.org/10.1080/10255842.2020.1793965
10. Wu, J.Z., Welcome, D.E., Krajnak, K., Dong, R.G.: Finite element analysis of the penetrations of shear and normal vibrations into the soft tissues in a fingertip. Med. Eng. Phys. **29**, 718–727 (2007). https://doi.org/10.1016/j.medengphy.2006.07.005
11. Hou, Z., Li, Z., Fadiji, T., Fu, J.: Soft grasping mechanism of human fingers for tomato-picking bionic robots. Comput. Electron. Agric. **182**, 106010 (2021). https://doi.org/10.1016/j.compag.2021.106010
12. Wu, J.Z., Welcome, D.E., Mcdowell, T.W., Xu, X.S., Dong, R.G.: Modeling of the interaction between grip force and vibration transmissibility of a finger. Med. Eng. Phys. **45**, 61–70 (2017). https://doi.org/10.1016/j.medengphy.2017.04.008
13. He, Y., Liu, H., Himeno, R., Sunaga, J., Kakusho, N., Yokota, H.: Finite element analysis of blood flow and heat transfer in an image-based human finger. Comput. Biol. Med. **38**, 555–562 (2008). https://doi.org/10.1016/j.compbiomed.2008.02.002
14. Askari, E., Flores, P., Dabirrahmani, D., Appleyard, R.: Nonlinear vibration and dynamics of ceramic on ceramic artificial hip joints: a spatial multibody modelling. Nonlinear Dyn. **76**, 1365–1377 (2014). https://doi.org/10.1007/s11071-013-1215-y
15. Machado, M., et al.: Development of a planar multibody model of the human knee joint. Nonlinear Dyn. **60**, 459–478 (2009). https://doi.org/10.1007/s11071-009-9608-7
16. Machado, M., Flores, P., Ambrosio, J., Completo, A.: Influence of the contact model on the dynamic response of the human knee joint. Proc. Inst. Mech. Eng. Part K J. Multi-body Dyn. **225**, 344–358 (2011). https://doi.org/10.1177/1464419311413988
17. Borucka, A., Ciszkiewicz, A.: A planar model of an ankle joint with optimized material parameters and Hertzian contact pairs. Materials (Basel) **12**, 2621 (2019)

18. Ciszkiewicz, A.: Analyzing uncertainty of an ankle joint model with genetic algorithm. Materials (Basel) **13**, 1175 (2020). https://doi.org/10.3390/ma13051175
19. Bajuri, M.N., Kadir, M.R.A., Raman, M.M., Kamarul, T.: Mechanical and functional assessment of the wrist affected by rheumatoid arthritis: a finite element analysis. Med. Eng. Phys. **34**, 1294–1302 (2012). https://doi.org/10.1016/j.medengphy.2011.12.020
20. Wei, F., Fong, D.T.P., Chan, K.M., Haut, R.C.: Estimation of ligament strains and joint moments in the ankle during a supination sprain injury. Comput. Methods Biomech. Biomed. Eng. **18**, 243–248 (2015). https://doi.org/10.1080/10255842.2013.792809
21. Draper, C.E., et al.: Is cartilage thickness different in young subjects with and without patellofemoral pain? Osteoarthritis Cartilage **14**, 931–937 (2006). https://doi.org/10.1016/j.joca.2006.03.006
22. Shepherd, D.E.T., Seedhom, B.B.: Thickness of human articular cartilage in joints of the lower limb. Ann. Rheum. Dis. **58**, 27–34 (1999). https://doi.org/10.1136/ard.58.1.27
23. Foumani, M., Strackee, S.D., Van De Giessen, M., Jonges, R., Blankevoort, L., Streekstra, G.J.: In-vivo dynamic and static three-dimensional joint space distance maps for assessment of cartilage thickness in the radiocarpal joint. Clin. Biomech. **28**, 151–156 (2013). https://doi.org/10.1016/j.clinbiomech.2012.11.005
24. Various: least squares circle. https://scipy-cookbook.readthedocs.io/items/Least_Squares_Circle.html

Is Optimal Velocity Constant During Running?

Ryszard Maroński(✉)

Institute of Aeronautics and Applied Mechanics, Warsaw University of Technology,
ul. Nowowiejska 24, 00-665 Warsaw, Poland
maron@meil.pw.edu.pl

Abstract. The minimum-time competitive run is reconsidered in the paper. The problem is formulated and solved in optimal control (calculus of variations). A non-classical method of Miele is applied – the method of extremization of line integrals by Green's theorem. This method gives necessary and sufficient conditions of optimality. Two models of the energy conversions in competitor's body are considered: the model of Keller and the model of Bonnans and Aftalion. It is shown that the optimal race may be broken into three phases: acceleration, cruise (along so-called singular arc), and final slowing down. The fundamental finding is that the optimal cruising velocity is constant for Keller's model but it decreases for Bonnans and Aftalion's model.

Keywords: Minimum-time running · Singular arc

1 Introduction

The paper is devoted to investigation of optimal strategies during competitive running applying optimal control (calculus of variations). A pioneering mathematical work is that of Keller [1, 2]. He applied calculus of variations. From his investigations it follows that the distance to be covered may be broken into three sections:

1) initial acceleration where competitor uses his maximal abilities,
2) cruise with constant velocity in the middle part of the distance along so-called singular arc,
3) negative kick at the end, where the runner's velocity decreases.

For distances not longer than 291 m, where the energy reserves are not depleted, only the first section appears. On the other hand, observations of pacing strategies show that for 400 m and 800 m running events the races are always won by people who run the first half of the race faster than the second half. It is not true for shorter races, or for longer, where the second half of the race is faster. The constant-speed pacing strategy is observed for races of a mile or longer [3].

The questions may arise: 1) is the pacing strategy optimal during real races? or 2) are the models of running used in optimal control suitable for the process under consideration?

A. Hadamus et al. (Eds.): BIOMECHANICS 2021, LNNS 328, pp. 110–122, 2022.
https://doi.org/10.1007/978-3-030-86297-8_10

The models base on two ordinary differential equations. The first one derives from Newton's second law of motion. This equation is generally accepted. Different models of resistive force are used: a linear function of velocity or proportional to the square velocity. The second differential equation describes the energy conversions in runner's body and this is the area where different models may be employed. A progress has been made in this area the last years. Bonnans and Aftalion [4] propose more realistic model that relies on the assumption of variable oxygen uptake comparing with Keller's model of constant oxygen uptake. Aftalion [5] found the optimal solution for such a model along so-called singular arc where the optimal cruising velocity decreases with the time. The applied reasoning is very sophisticated, however. In this paper, the same problem for Bonnans and Aftalion's model is reconsidered using the method of Miele [6] – the method of extremization of line integrals by Green's theorem. The method gives a solution in the energy-velocity plane, it gives necessary and sufficient conditions of optimality including singular arcs of linear type, and it gives a clear graphical interpretation. It is easy to consider the inequality constraints imposed on the state variables and that may generate problems in the classical approach. The difficulty with application of the method is such that not always it is possible to perform the problem in the form required by the theory – as a line integral of two variables.

2 Mathematical Models

Many authors have considered the problem of constant velocity in sports events. Keller [1, 2] using calculus of variations proves that the runner's velocity is constant during distance running in the middle part of the race (cruise), omitting the starting phase and the finish. Maroński [7] shows that the optimal velocity is constant for cycling even if the slope angle of the trace varies with the distance. Such finding is not true for the race with variable wind heading. In the paper of Maroński and Samoraj [8] it is shown that for Keller's model of running, but for variable slope of the trace, the optimal velocity is constant during the cruise. As it is mentioned above, such results do not agree with observations of real events [3]. Bonnans and Aftalion [4] propose an improved model of metabolism where the oxygen uptake is not constant as in Keller's model, but it depends on the energy reserves staying at competitor's disposal. The current paper bases on this model.

Consider Keller's model of running. The racer is modelled as a particle of the mass m – his mass centre. The competitor moves on a linear horizontal track. The vertical displacements of his mass centre resulting from the cyclic nature of the stride pattern and the displacement at the start are omitted.

The first equation relies on Newton's second law (Fig. 1)

$$\frac{dv}{dt} = f(t) - \frac{v}{\tau}, \tag{1}$$

where all quantities refer to the unit mass of the runner, $v(t)$ is the instantaneous runner's velocity, $f(t)$ is the variable propulsive force (a control variable in optimal control), and v/τ is the resistive force linearly depending on the velocity v (τ is a constant coefficient).

The second equation describes the energy conversions in the competitor's body

$$\frac{de}{dt} = \sigma - f\, v\,, \tag{2}$$

where: e is the available energy per unit mass, σ is a constant energy flow rate (energy equivalent per unit mass of $V\dot{O}2$), $f\,v$ is the actual mechanical power per unit mass used by the athlete to overcome the inertia and the resistance of motion.

Both ordinary differential Eqs. (1) and (2) should be supplemented by the initial conditions at the start to the race:

$$v(0) = 0\,, \qquad e(0) = e_0\,. \tag{3}$$

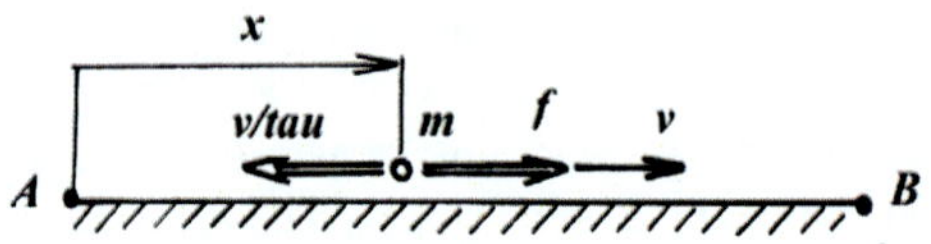

Fig. 1. The particle model of the competitor of the mass m. The forces exerted on his body per unit mass are: f – propulsive force (control variable), v/τ – resistive force. Furthermore: v - velocity of the runner, τ - constant coefficient, x is the actual covered distance, symbols A and B stand for initial and final points.

The hydraulic analogy of the energy flow model is given in Fig. 2. There are two containers representing different types of human energy. The upper vessel on the left contains a fluid representing an aerobic energy. The aerobic energy container is of infinite capacity. The fluid flows through the pipe to the lower container with a constant flow rate σ (energy equivalent per unit mass of $V\dot{O}2$). The energy flow rate does not depend on the fluid levels in both containers. The lower vessel contains the fluid representing the anaerobic energy e_0 at the beginning. Symbol e(t) denotes the energy per unit mass at a given instant of time. Bonnans and Aftalion [4] identify the energy e(t) as an anaerobic energy. It is equal to anaerobic energy at the beginning of the process. It is a mixture of both energies (aerobic and anaerobic) later; therefore it is regarded as the available energy in this paper. The difference (e_0 – e(t)) may be regarded as the accumulated oxygen deficit (*AOD*). There is another pipe at the bottom of the lower container. The energy flow rate ($f\,v$) may be regulated by a tap and adjusted to the actual conditions. The algorithm of regulation is not known at that moment and it should be found. Further details of the hydraulic analogy of the energy flow one can find in Bonnans and Aftalion [4].

From the point of view of optimal control Eqs. (1) and (2) are the state equations, v and e are the state variables, f is the control variable. Two inequality constraints should be satisfied. The first one:

$$0 \le f(t) \le f_{\max}\,, \tag{4}$$

where f_{max} is the maximal propulsive force per unit mass depending on the abilities of the runner, and the second one:

$$e(t) \ge 0\,, \tag{5}$$

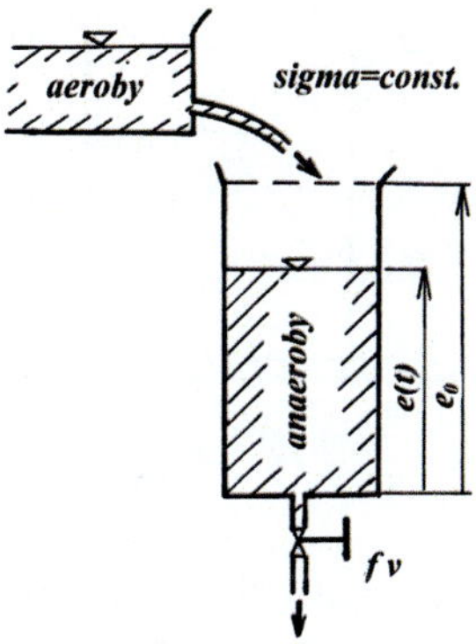

Fig. 2. Keller's model of the power balance. Symbol $e(t)$ - energy per unit mass, e_0 – initial energy, σ energy flow rate from the aerobic container to the anaerobic one, here $\sigma = \text{const.}$, $f\,v$ – power per unit mass, f – propulsive force per unit mass (control variable), v – velocity.

which means that the reserves of energy per unit mass cannot be negative during the race – the lower container may be filled with or empty.

The distance D to be covered is given, therefore:

$$D = \int_0^T v\,dt\,, \tag{6}$$

where: T is the time of the event.

The problem may be formulated in the following manner. The runner should vary his speed $v(t)$ during the race over a given distance D to minimize the time of the event T

$$T = \int_0^T dt\,. \tag{7}$$

The state Eqs. (1) and (2), the initial conditions (3), the inequality constraints (4), (5) and the isoperimetric constraint (6) should be satisfied. Aftalion considers an equivalent formulation of the problem: maximization of the distance D for the given time T of the event [5].

Now, consider another model of the energy flow in competitor's body [5] that is given in Fig. 3.

Here the fluid representing the energy per unit mass $e(t)$ is contained in communicating vessels. The left vessel contains the aerobic energy of infinite capacity. The right vessel contains the anaerobic energy at the beginning e_0. A pipe of finite diameter connects both vessels; therefore the fluid flow rate $\sigma(e)$ is limited and depending on the fluid level in the right container. This pipe is placed at the bottom of the right container. At the beginning of the process the fluid level is the same in both containers (the competitor is rested) – the fluid does not flow through the bottom pipe, $\sigma = 0$. If the right vessel is empty the fluid flows with maximal intensity $\overline{\sigma}$ from left container to the right one.

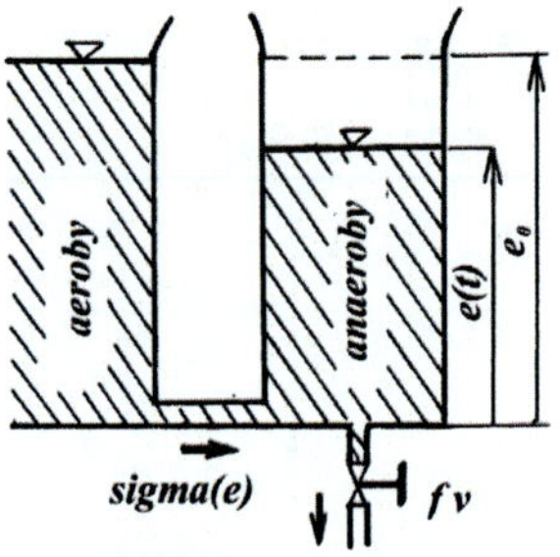

Fig. 3. Aftalion's model of the power balance. Symbol $e(t)$ - energy per unit mass, e_0 – initial energy, $\sigma(e)$ energy flow rate from the aerobic container to the anaerobic one, here σ depends on the fluid levels in both containers, $f\,v$ – power per unit mass, f – propulsive force per unit mass (control variable), v – velocity.

The model assumes that $\sigma(e)$ is a linear function of the energy per unit mass e and it is proportional to the accumulated oxygen deficit AOD = e_0-e:

$$\sigma(e) = \overline{\sigma}\left(1 - \frac{e}{e_0}\right), \tag{8}$$

where $\overline{\sigma}$ is the maximal value of σ. Now, Eq. (2) describing the energy conversions takes the form:

$$\frac{de}{dt} = \sigma(e) - f\,v\,. \tag{9}$$

Here, the constant σ from Keller's model is replaced by the function σ(e) given by Eq. (8). Further details of the model and the problem formulation are the same.

3 Method

3.1 Method of Miele

The method of Miele [6] was developed in the study of trajectories of high-speed aircraft and missiles, which could not be handled by conventional methods of performance analysis at the beginning of sixties. Here, a particular class of variational problems is considered, where the functional form to be extremized and the possible isoperimetric constraint are linear in the derivative of the unknown function $y(x)$.

Consider a functional linear in the derivative y':

$$J = \int_A^B \left[\phi(x,y) + y'\,\psi(x,y)\right] dx = \int_A^B \left[\phi(x,y)\,dx + \psi(x,y)\,dy\right] \tag{10}$$

that is minimized. Symbols φ and ψ denote known functions of two arguments (x,y). Symbols A and B stand for initial and final points. Assume that the process under consideration is represented by a curve joining points A and B in the (x,y)-plane (Fig. 4).

We assume that all solutions are within or along the border of the admissible domain represented by a region limited by a closed curve $\varepsilon(x,y) = 0$, and the points A and B also belong to this curve. Now, we can determine the difference in values of the integral (10) for two arbitrarily taken curves AQB and APB.

$$\Delta J = J_{AQB} - J_{APB} = \int_{AQB} (\phi\, dx + \psi\, dy) - \int_{APB} (\phi\, dx + \psi\, dy) = \oint_{AQBPA} (\phi\, dx + \psi\, dy) \tag{11}$$

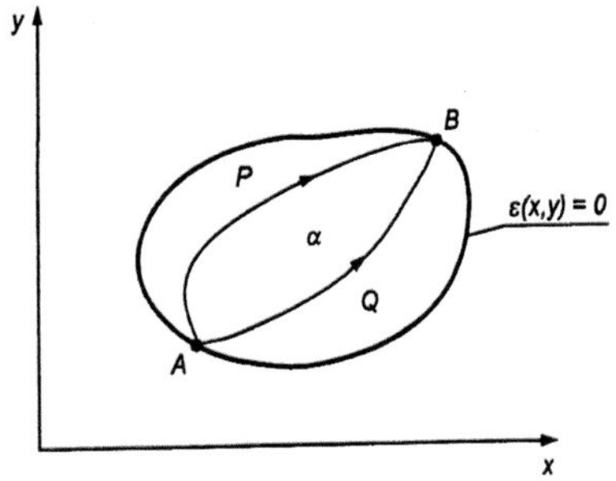

Fig. 4. The admissible domain is an area within the closed curve $\varepsilon(x,y) = 0$. Curves AQB and APB represent different admissible strategies.

Employing Green's theorem we can transform the cyclic integral (11) into a surface integral:

$$\Delta J = \iint_{\alpha} \omega(x, y)\, dx\, dy \tag{12}$$

where α represents a region limited by these two curves and ω is so-called fundamental function. It takes the form:

$$\omega(x, y) = \frac{\partial \psi}{\partial x} - \frac{\partial \phi}{\partial y} \tag{13}$$

Three cases are possible:

a) Fundamental function ω is identically equal to zero. That means $\Delta J = 0$, and the functional is independent of the curve in the (x,y)-plane – the process is irrespective of the strategy.
b) Fundamental function ω has the same sign, for example $\omega > 0$. For such the case $\Delta J > 0$ or $J_{AQB} > J_{APB}$. It means that every curve to the left gives smaller value of the functional. In the limit the minimizing curve belongs to the border of the admissible domain.
c) Fundamental function ω changes its sign. It means that there is a curve along which $\omega = 0$ that divides admissible domain into two subregions, where $\omega > 0$ and $\omega <$

0 respectively. The optimal path contains subarcs along the border of admissible domain and along the curve $\omega = 0$. This last subarc refers to the singular arc in calculus of variations (Fig. 5).

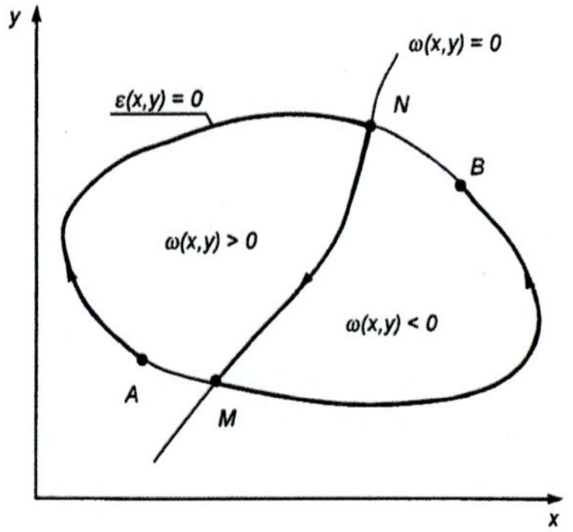

Fig. 5. The fundamental function ω changes its sign. The optimal solution contains two subarcs along the border of admissible domain AN, MB and the subarc within the admissible domain NM, where ω(x,y) = 0. The functional is minimized along the curve ANMB.

A modification of the previous problem occurs when the integral to be extremized (10) must satisfy also the linear isoperimetric constraint:

$$J = \int_A^B \left[\phi_1(x, y) + y' \, \psi_1(x, y)\right] dx = C , \tag{14}$$

where: φ_1 and ψ_1 denote known functions and C is a given constant. Now, the function ω must be replaced by the augmented function ω^* of the form:

$$\omega^* \, (x, y, \lambda) = \frac{\partial \psi^*}{\partial x} - \frac{\partial \phi^*}{\partial y} , \tag{15}$$

where

$$\phi^* = \phi + \lambda \phi_1 , \qquad \psi^* = \psi + \lambda \psi_1 . \tag{16}$$

The equation $\omega = 0$ of the simple problem must be replaced by the equation for augmented function $\omega^* = 0$. The equation for the isoperimetric problem represents a family of curves – one curve for each value of Lagrange's multiplier λ. The particular value of λ is to be determined from the given isoperimetric constraint (14).

3.2 Solution

Keller [1, 2] applies the classical methods of calculus of variations for the problem solution. Aftalion [5] uses the classical approach of optimal control applying Pontryagin's maximum principle with Hamiltonian linear in the control function – here the propulsive force $f(t)$. Aftalion's reasoning is very sophisticated. Maroński [10], for Keller's model of running, uses non-classical method of Miele. This method gives necessary and sufficient conditions of optimality including singular arcs of linear type. It gives clear graphical interpretation. This method is not popular because the problem under investigation should be reduced to the form required by the method of Miele. For Aftalion's model of running it is possible, however. The detailed description of the method is out of the scope of this paper. The details one can find in Maroński [9], pp. 24–29.

The minimum-time running problem for Aftalion's model [5] may be reduced to a line integral as it follows. Equations (1) and (2) may be expressed using x, the actual covered distance, instead of the time t, using the definition of the velocity:

$$v = \frac{dx}{dt} . \tag{17}$$

Now, the state equations take the forms:

$$\frac{dv}{dx} = \frac{f(x)}{v} - \frac{1}{\tau} , \tag{18}$$

$$\frac{de}{dx} = \frac{\sigma(e)}{v} - f(x) . \tag{19}$$

The isoperimetric constraint – the distance to be covered – is of the form:

$$D = \int_A^B dx , \tag{20}$$

and the time of the run to be minimized:

$$T = \int_A^B \frac{1}{v} dx , \tag{21}$$

where: A, B are the initial and final points.

Now, we can find the propulsive force $f(x)$ from Eq. (19) and put it into Eq. (18), then

$$\frac{dv}{dx} = \frac{\sigma(e)}{v^2} - \frac{1}{v}\frac{de}{dx} - \frac{1}{\tau} . \tag{22}$$

The differentials appearing in the integrals Eqs. (20) and (21) are as follows:

$$dx = \frac{\tau v}{a(e,v)} de + \frac{v^2 \tau}{a(e,v)} dv , \tag{23}$$

$$\frac{dx}{v} = \frac{\tau}{a(e,v)} de + \frac{v \tau}{a(e,v)} dv , \tag{24}$$

where:

$$a(e,v) = \sigma(e)\,\tau - v^2 \neq 0\,. \tag{25}$$

The isoperimetric constraint (20) and the time to be minimized (21) may be performed as line integrals

$$D = \int_A^B dx = \int_A^B \phi_1(e,v)\,de + \psi_1(e,v)\,dv\,, \tag{26}$$

where

$$\phi_1 = \frac{v\,\tau}{a(e,v)}\,, \qquad \psi_1 = \frac{v^2\,\tau}{a(e,v)}\,, \tag{27}$$

and

$$T = \int_A^B \frac{dx}{v} = \int_A^B \phi(e,v)\,de + \psi(e,v)\,dv\,, \tag{28}$$

where

$$\phi = \frac{\tau}{a(e,v)}\,, \qquad \psi = \frac{v\,\tau}{a(e,v)}\,. \tag{29}$$

According to the method of Miele [6], the augmented functions ϕ^* and ψ^* take the forms:

$$\phi^* = \phi + \lambda\,\phi_1 = \frac{\tau}{a(e,v)}\,(1 + \lambda\,v)\,, \tag{30}$$

$$\psi^* = \psi + \lambda\,\psi_1 = \frac{v\,\tau}{a(e,v)}\,(1 + \lambda\,v)\,, \tag{31}$$

where: λ is constant Lagrange's multiplier.

The augmented fundamental function is of the form

$$\omega^*(e,v,\lambda) = \frac{\partial\psi^*}{\partial e} - \frac{\partial\phi^*}{\partial v} = \frac{1}{a^2(e,v)}\left[-\tau\,v\,(1+\lambda\,v)\left(\tau\,\frac{d\sigma}{de} + 2\right) - \lambda\,\tau\,\left(\sigma(e)\,\tau - v^2\right)\right]. \tag{32}$$

Expression (32), after equating to zero, yields the optimal solution along so-called singular arc

$$\tau\,\frac{d\sigma}{de}\,(1 + \lambda\,v) + \frac{\lambda\,\sigma(e)\,\tau}{v} + \lambda\,v + 2 = 0\,. \tag{33}$$

It is an algebraic equation depending on the functions of the covered distance x: $e(x)$, $v(x)$, and constants. If there is a subarc, where

$$\sigma(e) = \overline{\sigma} = const.\,, \tag{34}$$

then $d\sigma/de = 0$ and the equation of the singular arc is an algebraic equation of a simpler form

$$\lambda\, v^2 + 2\, v + \lambda\, \overline{\sigma}\, \tau = 0\,. \tag{35}$$

The parameters λ, $\overline{\sigma}$, τ are constant therefore from Eq. (35) it follows that along such arc $v = v_{cruise} =$ const., and this result is significantly consistent with the results of Keller [1,2] and Maroński [10].

From Miele's method it follows, that the admissible domain is bordered by the curves obtained from integration of Eqs. (1) and (2) for $f(t) = f_{max}$ on the right, the equality $e(t) = 0$ on the left, and the equality (25), $a(e,v) = 0$, from the bottom (cf. Figs. 6 and 7). The distance of the race may be broken into three phases:

1. The early phase of the race (acceleration), where the competitor moves using his maximal propulsive force, $f(t) = f_{max}$. The velocity rapidly increases. For the distances short enough only such a phase occurs (sprints) – the energy reserves are not depleted. In the (e,v)-plane this phase represents the arc along the border of admissible domain joining the initial point A and the singular arc.
2. The middle phase of the race (cruise). Here the competitor moves employing his partial propulsive force, $0 < f(t) < f_{max}$. His velocity moderately varies with the time or even it is constant. In the (e,v)-plane this phase corresponds to the arc within the admissible domain. The arc extends from the border where $f(t) = f_{max}$, to the border where $e(t) = 0$. This section corresponds to the singular arc in calculus of variations or optimal control.
3. The finish with decreasing velocity (negative kick). The border of the admissible domain is reached. Here the energy from the anaerobic container is depleted, $e(t) = 0$. The energy is supplemented by the flow from aerobic container, but the intensity of the flow is relatively low. The propulsive force f(t) follows from Eqs. (2) or (9), $f = \sigma(e)/v$. The energy flow rate $\sigma(e)$ is too small to sustain the cruising velocity. The final point B may be reached along this curve.

Originally, the method of Miele was worked out for the given final point B (given final velocity v_B). This difficulty may be overcame in the manner described in the book of Maroński [9], where the problem with unspecified final velocity is considered as a sequence of the problems with fixed final velocity and minimizing the time of the event with respect to v_B.

4 Results

An example of the minimum-time run for constant σ over the distance $D = 400$ m using the method of Miele is given in Maroński [10]. Originally this problem solved Keller [1, 2] using the classical methods of calculus of variations. The data for computations were adjusted to the results at that time and they are: $e_0 = 2409.25$ m^2/s^2, $f_{max} = 12.2$ m/s^2, $\tau = 0.892$ s, $\sigma = 41.61$ m^2/s^3. The track is inclined to the horizontal and the inclination angle is 2^0. The admissible domain is given in Fig. 6 (undashed area). The optimal path consists of the subarcs: the acceleration with $f(t) = f_{max}$ on the right, the cruise with constant velocity in the middle, and the negative kick at the end. The details one can find in Maroński [9].

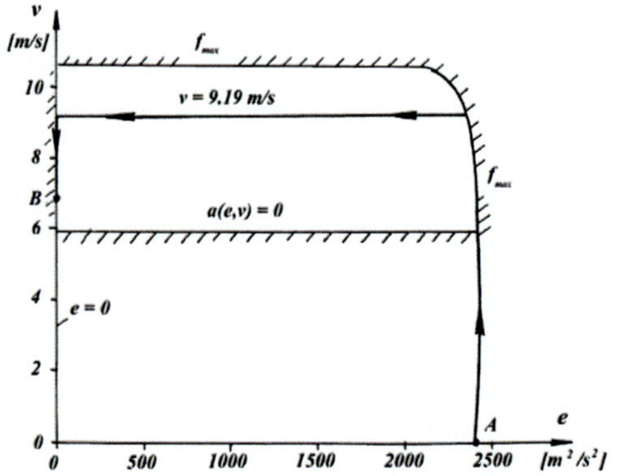

Fig. 6. The admissible domain (undashed area) for Keller's model of metabolism ($\sigma = const.$), the 400 m run and the slope angle of the track 2^0. The velocity is constant along the singular arc (in the middle).

The second example employs Aftalion's model of metabolism [5], where the energy flow rate from the aerobic container to the anaerobic one is not constant, but it depends on the energy according to Eq. (8). The data for computations are as follows (cf [5, 11]): e_0= 2409 m^2/s^2, f_{max}= 9 m/s^2, $\tau = 0.89$ s, $\overline{\sigma}$= 22 m^2/s^3. The track is horizontal. Here also the maximum propulsive force $f(t) = f_{max}$ is developed at the early phase of the race (acceleration). The optimal velocity slightly decreases along the singular arc then. The finish is with decreasing velocity along the subarc, where $e(t) = 0$. The difference, comparing with Keller's model, is along the singular arc, where the cruising velocity is not constant. There is the family of singular arcs for different values of Lagrange's multiplier in Fig. 7. The singular arc for $\lambda = -0.19$ s/m is on the outside of the admissible domain – it is not active. The singular arc for $\lambda = -0.23$ s/m does not reach the left

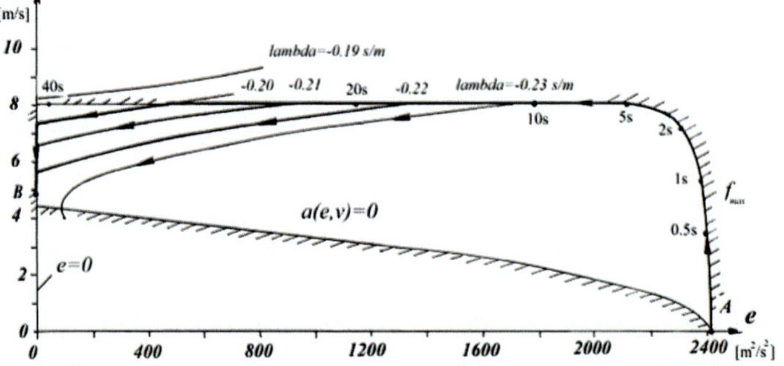

Fig. 7. The admissible domain (undashed area) for Aftalion's model of metabolism ($\sigma = \sigma(e)$), and the run over different distances represented by different values of constant Lagrange's multiplier λ. Here the velocity slightly decreases along the singular arc (in the middle).

border of the admissible domain, $e(t) = 0$. It cuts the bottom border of the admissible domain, where $a(e,v) = 0$. The method of Miele disappoints for such the case.

5 Discussion

The minimum-time competitive run is reconsidered in the paper. Such a problem may be formulated and solved in optimal control (calculus of variations). The question formulated at the beginning is this: What is the runner's optimal velocity profile along the distance minimizing the time of the event? Another question may be put: Is such a section of the race where the optimal velocity is constant? Such a problem solved Keller [1, 2]. He employs relatively simple model of motion basing on Newton's second law and an equation of the power balance. In his model the energy flow rate from the aerobic container to the anaerobic one is assumed to be constant. The result of Keller's considerations is that the race may be broken into three phases:

- acceleration,
- cruise with constant competitor's velocity,
- and final slowing down.

Such the result does not fit well the results observed during real events. Aftalion [5] reconsidered the problem improving the model of the power balance in the competitor's body. Her model assumes that the energy flow rate between the aerobic container and the anaerobic one is not further constant, but it depends on the energy level in the anaerobic vessel (cf. Eq. (8)). For the problem solution she applies classical Pontryagin's maximum principle with the Hamiltonian linear in the control function – here the propulsive force $f(t)$. Such an approach is very sophisticated and it gives only necessary conditions of optimality. In the present paper the non-classical method of Miele is applied – the method of extremization of line integrals by Green's theorem. This method gives necessary and sufficient conditions of optimality and it gives a simple graphical interpretation. The difficulty is that not all problems may be performed in the form required by this method – as a line integral of two variables. It was the reason that the minimum-time problem for running over variable slope trace was solved using a numerical method - direct Chebyshev's pseudospectral method [8]. For Keller's and Aftalion's models of running such transformation is possible, however. In this paper it is proved that the optimal path consists of three sections: acceleration, cruise and final slowing down. The optimal cruising velocity is not constant, as in Keller's model – it slightly decreases. The equation of the singular arc is an algebraic Eq. (33) and not an ordinary differential equation (cf. [5], Eq. (17)). The qualitative result is the same – the optimal velocity decreases along the singular arc (during the cruise). The answer for the question put in the title of the paper is: the optimal cruising velocity may be constant or decreasing. It depends on the model of the power balance used, and not on the distance to be covered. According to Reardon [3], it seems that the model used by Aftalion [5] is better for middle distance running comparing with Keller's model.

References

1. Keller, J.B.: A theory of competitive running. Phys. Today **26**, 43–47 (1973)
2. Keller, J.B.: Optimal velocity in a race. Am. Math. Monthly **81**, 474–480 (1974)
3. Reardon, J.: Optimal pacing for running 400 m and 800 m track races. Am. J. Phys. **81**, 428–435 (2013)
4. Bonnans, J.F., Aftalion, A.: Optimization of running strategies based on anaerobic energy and variations of velocity. INRIA Research Report no. 8344 – Août 2013, 24p. https://hal.inria.fr/hal-00851182v1
5. Aftalion, A.: How to run 100 meters. SIAM J. Appl. Math. **77**, 1320–1334 (2017)
6. Miele, A.: Extremization of linear integrals by Green's theorem. In: Litmann G. (ed.) Optimization Techniques with Application to Aerospace System, pp. 69–98. Academic Press, New York (1962)
7. Maroński, R.: On optimal velocity during cycling. J. Biomech. **27**, 205–213 (1994)
8. Maroński, R., Samoraj, P.: Optimal velocity in the race over variable slope trace. Acta Bioeng. Biomech. **17**, 149–153 (2015)
9. Maroński, R.: Strategie optymalne w mechanice lotu i biomechanice. Oficyna Wyd. Politechniki Warszawskiej 79–93 (2016) (in Polish)
10. Maroński, R.: Minimum-time running and swimming: an optimal control approach. J. Biomech. **29**, 245–249 (1996)
11. Aftalion, A., et al.: How to identify the physiological parameters and run the optimal race. Math. Action **7**, 1–10 (2016)

Characteristics of Nerve Roots Mechanical Properties Exposed to Uniaxial Stretching Tests

Monika Palmerska[1](✉), Agnieszka Mackiewicz[1], Tomasz Klekiel[1], Agnieszka Noszczyk-Nowak[2], and Romuald Będziński[1]

[1] University of Zielona Góra, Zielona Góra, Poland
20000884@stud.uz.zgora.pl, {a.mackiewicz,t.klekiel, r.bedzinski}@iimb.uz.zgora.pl

[2] Wrocław University of Environmental and Life Sciences, Wrocław, Poland
agnieszka.noszczyk-nowak@upwr.edu.pl

Abstract. The article presents the problem of the mechanical response of nerve roots under stretching conditions. Research into the functioning of the nervous system shows a relation between deformation, blood flow characteristics, and nerve root impulse transmission. These studies show that vascular hypofusion of peripheral nerves occurs at a deformation of 15%. It is also known that impulse conduction disturbances occur at 6% strains. This calculation shows that one of the causes of diseases of the nervous system may be excessive deformation of its structures. One such disease is radiculopathy. Understanding the mechanical response of tissues exposed to damaging conditions requires computations from complex constitutive models that cannot be solved analytically. Therefore, the properties of the tissue material are investigated using numerical approximation methods. For this purpose, it is necessary to experimentally study the properties of the material that allow to determine the parameters of the numerical models used. The aim of the research was to determine the material properties of nerve roots. The analysis was carried out in order to obtain data that would simulate the conditions of the development of nerve root diseases. Modelling was performed using the Finite Element Method (FEM). Mechanical properties were determined on the basis of uniaxial stretching of the nerve roots of the rabbit's lumbar spine on a testing machine. The tests showed that the rabbit's nerve root strength is 0.9 ± 0.53 MPa, the relative deformation is $12.73 \pm 4.03\%$, and the Young's modulus is 2.53 ± 1.00 MPa. The results show the range of strength and stiffness values for various species of animals and humans. The conducted research has shown that the sample preparation procedure has an important influence on the obtained values.

Keywords: Nerve roots · Radiculopathy · Uniaxial stretching tests

1 Introduction

Radiculopathy is a condition caused by chemical or mechanical factors that destroy the nerve roots. It consists of disturbances in the dorsal roots or dorsal root ganglia and

A. Hadamus et al. (Eds.): BIOMECHANICS 2021, LNNS 328, pp. 123–131, 2022.
https://doi.org/10.1007/978-3-030-86297-8_11

is revealed by various types of pain. Nerve roots may be exposed to excessive violent stresses during traumatic childbirth, traffic accidents or sports accidents [12]. Root damage also results from iatrogenic causes derived from insufficient knowledge about the influence of external influences on disturbances in the functioning of the nervous system. Radiculopathy is accompanied by various types of pain and functional deficits that make life difficult for patients. In addition to compression, nerve roots can be exposed to tensile stress, for example by adhesions to surrounding structures due to inflammation. This phenomenon is often unnoticeable during follow-up examinations using magnetic resonance imaging (MRI), which makes proper diagnosis difficult [1, 2]. This disease is one of the most common causes of neuropathic pain [3]. They are diagnosed in up to 40–50% of patients with chronic back and shoulder pain [4], and just lumbar radiculopathy affects 3 to 5% of the general population [5]. The diagnosis of lumbar radiculopathy is an indication for surgery. Decompression and stabilization of the spine are effective methods of treating pain in patients with mechanical radiculopathy [6], but there are known cases of neurological deficits associated with abnormal position of the screw [7]. In the case of stabilization of the spine with mechanically fixed implants, the contact of the nerve root with the screw leads to inflammation, further to morphological hypertrophy and ultimately to disturbances in the transmission of nerve impulses. Even with a relatively small rupture of the bark, which has an average thickness of about 2 mm, root damage may occur [7, 8]. The studies conducted so far indicate that even vertebral injuries without overt perforation of the cortex may cause significant radiculopathy [7–10]. It is unclear whether, when and how screw-induced damage to the cortical part of the shaft should be assessed as a possible cause of radiculopathy [7].

The cause of physiological and structural changes in the nervous tissue exposed to excessive mechanical stress is i.a. ischemia of nervous structures. Nerve vascular hypofusion occurs at a strain of 15%, and tissue changes are seen at an elongation of between 4% and 50%. In order to determine the relationship between deformation and the functioning of the nervous system, impulse conduction disorders can be observed. Literature data show that impulse conduction disturbances occur from 6% of deformation [13, 16, 17]. It is important to understand the development of nerve pathology resulting from mechanical action to understand the relationship between deformations and the structural and functional reaction of nerves [13].

Increasing the pressure inside the nerve leads to a reduction in blood flow in the root, histological changes such as edema, electrophysiological changes, such as reduced conduction velocity and increased excitability of dorsal root ganglia [1]. There is a linear relationship between deformation and the appearance of axonal damage in the form of impaired axoplasmic transport, fiber tear, and primary axotomy. Usually, the explanation of the pathomechanisms of nerve root injuries is carried out by experiments conducted on the nerve roots of humans and animals. [11, 12, 18] and numerical research [24, 25].

Understanding the mechanical response of tissues exposed to damaging conditions requires computations from complex constitutive models that cannot be solved analytically due to the complexity of the equations. Therefore, the parameters of the tissue material are investigated using numerical approximation methods, e.g. simulating mechanical loads using the finite element method [22, 27, 28].

Khuyagbaatar B. et al. [26] developed a finite element model of the cervical spine together with the human spinal cord and nerve roots. They analyzed the stresses and displacements of the spinal cord and nerve roots in three different positions: straight, lordosis and kyphosis. The studies compared models before and after C3–C7 laminectomy in the case of posterior longitudinal ligament ossification. Research results indicate that, with the exception of lordosis, stresses in the nerve roots are higher after laminectomy. Researchers suggest that this may be the cause of postoperative C5 nerve palsy, which is one of the most common complications after cervical spine surgery due to ossification of the posterior longitudinal ligament. Nerve roots were divided into external and intradural. The epidural nerve roots were assigned elastic material properties and the intradural nerve roots were given "nonlinear force-displacement relations". Lee N. et al. [21] created a two-dimensional numerical model of the L4–L5 intervertebral disc hernia with epidural fusion of nerve roots. Nerve structures have been assigned linear-elastic material properties.

The study investigated the mechanical properties of nerve roots subjected to the tensile loads. This explanation is important in understanding their structural and functional tolerance, preventing injury, and effectively treating radiculopathy. The aim of the research was to determine the material properties of nerve roots. The experiment was carried out on nerve roots taken from the lumbar region of the White New Zealand Rabbit. The obtained data from the tensile test were used to evaluate the mechanical properties of the samples and to determine and adjust the parameters of the material model [20].

2 Materials and Methods

The preparations were obtained as secondary material from New Zealand White Rabbits included in the urinary biomechanical research permit according by the Local Ethics Committee in Wroclaw (decision No. 1/2017).The lumbar spine of a rabbit was harvested and then stored in the BIOLASOL® (Biochefa, Sosnowiec, Poland) solution in a freezer at 5 °C for 6 h. BIOLASOL® is solution for transplantation‘s organs preservation and perfusion, preserves the tissue and does not change its mechanical properties. Paraspinal muscles and spinous processes were removed. Tissue samples from the spinal cord along with the spinal nerve were collected individually with a scalpel and tweezers just before the examination. To prevent drying out, the material was soaked in BIOLASOL® (Biochefa, Sosnowiec, Poland) solution. Tensile tests were performed at a velocity of 0.17 mm/s on a ZWICK Roell testing machine. This velocity value was chosen as typical value used to the quasi-static simulations. The most commonly used value about 0.01 s^{-1} was related to the average length of the samples. The tested tissues were properly fixed in order to minimize damage that could affect the obtained values of forces and deformations (Fig. 1).

In order to increase the probability that the specimen will not move in the fixture, a high roughness grid was used. Tensile testing was carried out under the conditions: air temperature equals 23.5 °C, humidity 60%. The environmental chamber was not used.

Fig. 1. View of the sample attached to the holders of the testing machine.

3 Results

The data obtained in the uniaxial tensile tests indicate that the strength of the nerve roots is 0.9 ± 0.53 MPa, the relative deformation is $12.73 \pm 4.03\%$, and the Young's modulus is 2.53 ± 1.00 MPa. The data from stretching 4 to 7 samples were used for the analyzes.

Table 1. Results of experiments.

No.	Max. stress [MPa]	Young modulus [MPa]	Max strain [%]	Velocity [mm/s]	Strain rate [s^{-1}]
1	0.4	1.11	11.26	0.17	0.1
2	1.64	2.82	9.52	0.17	0.1
3	0.74	2.7	11.5	0.17	0.1
4	0.83	3.47	18.63	0.17	0.1

Samples that slipped from their holders during the experiment were excluded. The obtained results are presented in Table 1.

4 Discussion

Kwan et al. [18] reported that human spinal nerve roots had a tensile strength of 0.17 ± 0.59 MPa and a final strain of $15.0 \pm 3.5\%$ at a rate of 0.17 mm/s. Sunderland and Bradley [19] reported that the range of maximum tensile stress and load was 3.9–29.4 MPa and 0.2–3.3 kg, respectively, in human roots of the third sacral nerve stretched at a rate of 1.27 mm/s.

Nashida et al. [11] conducted uniaxial stretching tests of pig nerve roots and root branches in order to determine their strength characteristics. The tests were carried out at different stretching rates: 0.01, 0.1, 1 and 10 s^{-1}. There has been a relatively constant relationship between endurance and nerve root deformation, despite changes in parameters. At each deformation rate, the tensile strength was approx. 0.7 MPa. The tensile strength of the root branches decreased with the increase in speed from about 2.3 MPa at the speed of 0.01 s^{-1} to 1.5 MPa at the speed of 10 s^{-1}.

Table 2 presents the results obtained in the conducted experiment with the data available in the literature. Singh A. et al. [12] performed stretching tests on rat nerve roots at two different stretching rates: 0.01 mm/s and 15 mm/s. Comparing the mechanical properties at both speeds revealed significant differences. When tensile at a speed of 0.01 mm/s, the maximum stress was 257.9 ± 111.3 kPa, and at a speed of 15 mm/s, it was 624.9 ± 306.8 kPa. There were no differences in the deformation values.

Table 2. Comparison of the obtained results with the literature data

Max stress [MPa]	Young modulus [MPa]	Max strain [%]	Velocity [mm/s]	Strain rate [s^{-1}]
Results of experiments				
0.9 ± 0.53	2.53 ± 1.00	12.73 ± 4.03	0.17	0.1
Nashida N. et al. [11]				
0.7	2.25	No data	0.1	0.01
0.7	2.1	No data	1	0.1
0.7	1.8	No data	10	1
0.7	1.6	No data	100	10
Singh A. et al. [12]				
0.258 ± 0.111	1.3 ± 0.8	29.0 ± 8.9	0.01	0.00033
0.625 ± 0.307	2.9 ± 1.5	30.8 ± 8.4	15	0.5
Sunderland S. et al. [19]				
3.9 ± 2.94	No data	9–16	1.27	0.0508
Kwan MK. et al. [18]				
1.71 ± 0.59	No data	15.0 ± 3.5	0.17	No data

The results presented in Table 2 indicate significant differentiation of the obtained mechanical properties of nerve roots. Literature data show discrepancy in terms of the conditions to which the samples were given during stretching. Important in the data analysis is also the fact that the tested roots were collected from various species of animals.

Figure 2 summarizes the values of the linear deformability modulus, taking into account the standard deviation of the own research and literature data. The Young's modulus range values obtained in the experiment is marked with red lines. The obtained

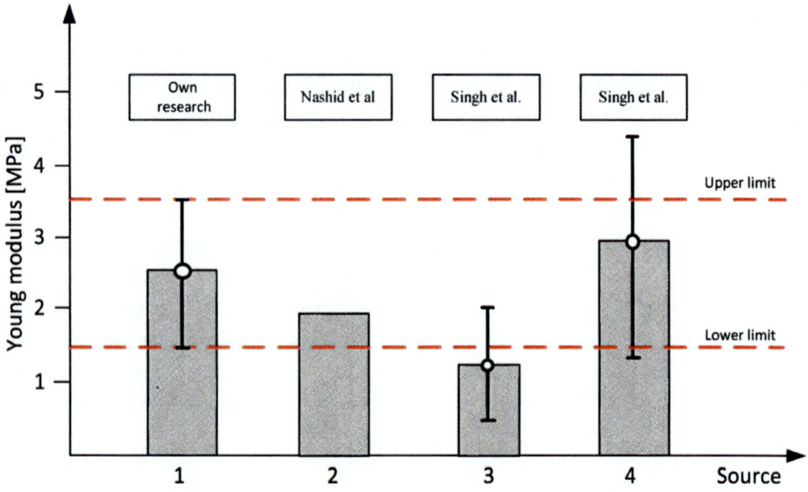

Fig. 2. Determination of the Young's modulus taking into account the standard deviation. 1 – own research, 2 – Nashid et al., 3 – Singh et al. At a deformation speed of 0.01 mm/s, 4 – Singh et al. With a deformation speed of 15 mm/s.

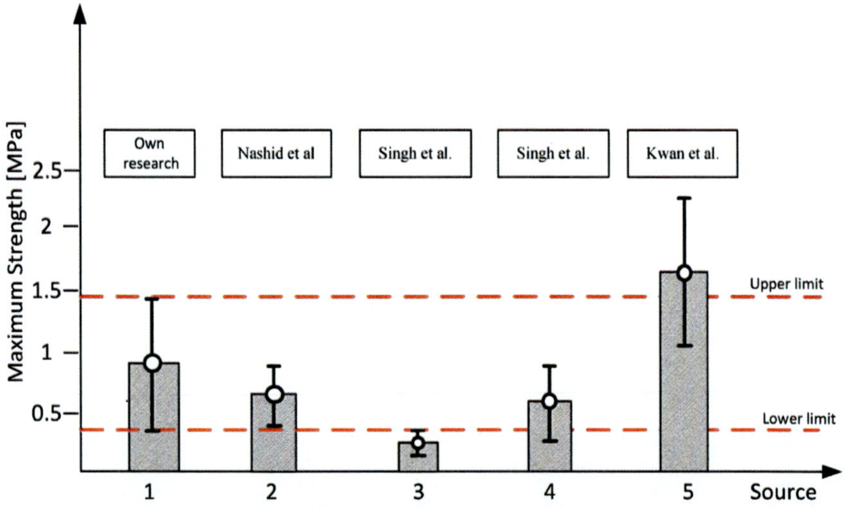

Fig. 3. The obtained maximum stresses taking into account the standard deviation. 1 – own research, 2 – Nashid et al., 3 – Singh et al. At a deformation speed of 0.01 mm/s, 4 – Singh et al. At a deformation speed of 15 mm/s, 5 – Kwan et al.

results partially overlap with the results of other researchers, but none of the data fully overlap with each other.

A similar situation can be seen in Fig. 3, where the results of maximum stresses are summarized. Nashida et al. carried out experiments to stretch the pig's nerve roots, Singh et al. rat, and Kwan et al. Stretched human's tissue. The results obtained by us fall within the ranges provided by other researchers who conducted tests at different strain rates. The results of the maximum strength resulting from the literature data range from 0.147 MPa to 2.3 MPa. Our results range from 0.37 MPa to 1.43 MPa.

5 Conclusion

The tensile strains achieved in presented research and in the combined literature data exceed 9%, which in clinical conditions would result in vascular hypofusion, conduction disturbances and tissue changes [13, 16, 17]. Ju et al. reported that blood perfusion in the sciatic nerve starts drops sharply at a pressure of 30.5 mmHg, i.e. 0.004 MPa [23]. As a result of uniaxial tensile tests to test material strength, however, tissue continuity is broken, and thus tissue destruction, which in clinical conditions would result in impairment of its functioning. The obtained data do not provide an answer to what are the stresses in the nerve roots that do not cause disturbances or damage.

Thanks to the uniaxial nerve root stretching experiments, the data were obtained, according to which the nerve root strength of the rabbit's lumbar spine is 0.9 ± 0.53 MPa, the relative deformation is $12.73 \pm 4.03\%$, and the Young's modulus is 2.53 ± 1.00 MPa Young's modulus. The obtained results are similar to the data available in the literature. The discrepancies may be due to the fact that the experiments were carried out on the nerve roots of different species of animals and different strain rates. The literature data show that ischemic changes, resulting from the mechanical stress on nerves, contribute to disturbances in the conduction of nerve impulses. Therefore, monitoring of nerve conduction during the experiments would also be important for understanding the development of radiculopathy.

Acknowledgments. The protocol of the study was approved by the Local Ethics Committee in Wroclaw(decision No. 1/2017) for grant number DEC-2016/21/B/ST8/01972 funded by the National Science Center Poland.

Funding. This research was funded by the Ministry of Science and Higher Education of Poland, grant number 003/RID/2018/19.

References

1. Berthelot, J., Laredo, J.D., Darrieutort-Laffite, C., Maugaras, Y.: Stretching of roots contributes to the pathophysiology of radiculopathies. Joint Bone Spine **85**, 41–45 (2018)
2. Lin, J.H., Chiang, Y.H., Chen, C.C.: Lumbar radiculopathy and its neurobiological basis. World J. Anesthesiol. **3**(2), 162–173 (2014)
3. Bouhassira, D.: Neuropathic pain: definition, assessment and epidemiology. Rev. Neurol. **175**(1–2), 16–25 (2019)
4. Harifi, G., et al.: Prevalence of chronic pain with neuropathic characteristics in the general population. Pain Med. **14**(2), 287–292 (2013)
5. Barry, J.A., Elia, C., Saini, H.S., Miulli, D.E.: A review of lumbar radiculopathy, diagnosis, and treatment. Cureus **11**(10), e5934 (2019)

6. Moliterno, J., Vesselis, C.A., Hershley, M.A., Lis, E., Laufer, I., Bilsky, M.H.: Improvement in pain after lumbar surgery in cancer patients with mechanical radiculopathy. Spine J. **14**(10), 2434–2439 (2014)
7. Woo, E.J., Dicuccio, M.N.: Clinically significant pedicle screw malposition is an underestimated cause of radiculopathy. Spine J. **18**(7), 1166–1171 (2018)
8. Amato, V., Giannachi, L., Irace, C., Corona, C.: Accuracy of pedicle screw placement in the lumbosacral spine using conventional technique: computed tomography postoperative assessment in 102 consecutive patients. J. Neurosurg. Spine **12**, 306–313 (2010)
9. Schizas, C., Michel, J., Kosmopoulos, V., Theumann, N.: Computer tomography assessment of pedicle screw insertion in percutaneous posteriori transpedicular stabilization. Eur. Spine J. **16**, 613–617 (2017)
10. Fu, T.S., et al.: Computer-assisted fluoroscopic navigation of pedicle screw insertion: an in vivo feasibility study. Acta Orthop. Scand. **75**, 730–735 (2004)
11. Nashida, N., Kanchiku, T., Ohgi, J., Ichikara, K., Chen, X., Taguchi, T.: Mechanical properties of nerve roots and rami radiculares isolated from fresh pig spinal cords. Neural Regen. Res. **10**(11), 1869–1873 (2015)
12. Singh, A., Lu, Y., Chen, C., Cavanaugh, J.M.: Mechanical properties of spinal nerve roots subjected to tension at different strain rates. J. Biomech. **39**, 1669–2167 (2006)
13. Kitab, S.A., Miele, V.J., Lavelle, W.F., Benzel, E.C.: Pathoanatomic basis for stretch-induced lumbar nerve root injury with a review of the literature. Neurosurgery **65**(1), 161–168 (2009)
14. Abdulla, F.A., Smith, P.A.: Axotomy- and autotomy-induced changes in the excitability of rat dorsal root ganglion neurons. J. Neurophysiol. **85**(2), 630–643 (2001)
15. Bain, A.C., Raghupathi, R., Meaney, D.F.: Dynamic stretch correlates to both morphological abnormalities and electrophysiological impairment in a model of traumatic axonal injury. J. Neurotrauma **18**(5), 499–511 (2001)
16. Smith, D.H., Wolf, J.A., Lusardi, T.A., Lee, V.M.Y., Meaney, D.F.: High tolerance and delayed elastic response of cultured axons to dynamic stretch injury. J. Neurosci. **19**(11), 4263–4269 (1999)
17. Wall, E.J., Massie, J.B., Kwan, M.K., Rydevik, B.L., Myers, R.R., Garfin, S.R.: Esperimental stretch neuropathy. J. Bone Joint Surg. **74-B**, 126–129 (1992)
18. Kwan, M.K., Rydevik, B., Myers, R.R., Triggs, K., Woo, S.L.Y., Garin, S.: Biomechanical and histological assessment of human lumbosacral spinal nerve roots. Trans. Orthop. Res. Soc. **14**, 348 (1989)
19. Sunderland, S., Bradley, K.C.: Stress–strain phenomena in human spinal nerve roots. Brain **84**(1), 120–124 (1961)
20. Klekiel, T., Malesa, P., Sławiński, G., Będziński, R.: Protection capabilities of the ankle joint against the consequences of impact load. In: AIP Conference Proceedings. AIP Publishing LLC, p. 020086.2 (2019)
21. Lee, N., et al.: Finite element analysis of the effect of epidural adhesions. Pain Physician **19**(5), 787–793 (2016)
22. Palmerska, M., Klekiel, T.: Measurement of dynamic properties of animal liver to identify the material model. In: Korbicz, J., Maniewski, R., Patan, K., Kowal, M. (eds.) PCBEE 2019. AISC, vol. 1033, pp. 256–264. Springer, Cham (2020). https://doi.org/10.1007/978-3-030-29885-2_23
23. Ju, M.S., Lin, C.C.K., Fan, J.L., Chen, R.J.: Transverse elasticity and blood perfusion of sciatic nerves under in situ circular compression. J. Biomech. **39**(1), 97–102 (2006)
24. Kuryło, P., Klekiel, T., Pruszyński, P.: A mechanical study of novel additive manufactured modular mandible fracture fixation plates – preliminary study with finite element analysis. Injury **51**(7), 1527–1535 (2020)

25. Arkusz, K., Klekiel, T., Sławiński, G., Będziński, R.: Pelvic vertical shear fractures: the damping properties of ligaments depending on the velocity of vertical impact load. In: AIP Conference Proceedings. AIP Publishing LLC, p. 020077 (2019)
26. Khuyagbaatar, B., Kim, K., Park, W.M., Kim, Y.H.: Biomechanical investigation of post-operative C5 palsy due to ossification of the posterior longitudinal ligament in different types of cervical spinal alignment. J. Biomech. **57**, 54–61 (2017)
27. Klekiel, T., et al.: Novel design of sodium alginate based absorbable stent for the use in urethral stricture disease. J. Mater. Res. Technol. **9**, 9004–9015 (2020)
28. Mackiewicz, A.G., Klekiel, T., Kurowiak, J., Piasecki, T., Bedziński, R.: Determination of stent load conditions in New Zealand white rabbit urethra. J. Funct. Biomater. **11**(4), 70 (2020)

The Influence of Polyurethane Double-J Stent of Various Diameters on Urological Encrustation

Kamila Pasik(✉)

Department of Biomedical Engineering, Faculty of Mechanical Engineering, University of Zielona Góra, Licealna 9 Street, 65-417 Zielona Góra, Poland
k.pasik@iimb.uz.zgora.pl
http://www.kib.uz.zgora.pl

Abstract. Double-J ureteral stents (DJ stents) are commonly used in urology. The most troublesome difficulty in their application is stent encrustation, stent breakage, and recurrent urinary tract infections. This work aimed was to determine the causes of a double-J stent complication.

The analysis of DJ stents included microscopic analysis, mechanical strength testing, and surface roughness measurements. The analysis was performed on the brand new ureteral stents and ureteral stents implanted after the ureterorenoscopic-lithotripsy (URSL) to treat calcium oxalate stone. Performed analysis of the polyurethane DJ stents concentrated on elaborating the influence of ureteral stent diameter (1.00; 1.33; 1.66 mm) on the affinity to stent encrustation and/or breakage.

The comparison of DJ stents with different diameters confirmed that the risk of encrustation increases with the lower diameter of the ureteral stent. Remaining post-URSL kidney stone fragments deposited on the surface of the DJ stents formed a multilayer structure creating a risk of obstruction or blockage of ureteral stents and block the urine flow. Additionally, a decrease in mechanical strength of the DJ stent related to the implantation time was determined. DJ stent implantation caused an increase in surface roughness.

The performed analysis indicates the need for further exploration of the biomaterials used in DJ stents and modification of their surfaces. This may allow to eliminate the phenomenon of encrustation of urinary stone fragments and NaCl crystals on the surface of implanted stents.

Keywords: Urolithiasis · Ureteral stent · Encrustation · Scanning electron microscopy · Mechanical strength · Ureterorenoscopic-lithotripsy procedure · Surface roughness

1 Introduction

Urinary tract diseases are affecting an increasing proportion of the human population every year. In many countries, the percentage of people suffering from kidney disease reaches 10–12%. Urological diseases are seen as problematic because they are usually not the first diagnosis due to the comorbidities disease such as diabetes. The most common

A. Hadamus et al. (Eds.): BIOMECHANICS 2021, LNNS 328, pp. 132–144, 2022.
https://doi.org/10.1007/978-3-030-86297-8_12

renal disease that affects adult and pediatric patients is urolithiasis [1]. About 20% of the population suffers from that pathology [2].

The treatment of urolithiasis is a combination of many factors, such as pharmacotherapy, proper diet, ultrasound-guided stone disruption, and ultimately surgery. The most common surgical procedure used to treat urolithiasis is URSL (ureterorenoscopic-lithotripsy) due to its high global efficiency (92.7–95%) [3]. URSL is known as a mini-invasive procedure and involves crushing and removing stone deposits located in the patient's urinary tract using an endoscope through the urethra [4].

A lot of urological pathologies are treated with the help of ureteral stents. An absolute indication for stent insertions are ureteric injures, relief of obstructed pyelonephritis and bilateral ureteral obstructions. Implantation of ureteral stents is associated with a risk of infection due to bacterial accumulation on their surface. The main problems with stent implantation are stent migration, occlusion, encrustation and biofilm creation, leading to stone formation.

After surgery, the implantation of a double-J stent is usually needed. A DJ stents in the ureter is needed to improve the urine flow from the kidney to the bladder. After the laser treatment, stone fragments should flow along the ureter and leave the patient's body with the urine. However, in practice, the stone fragments flowing down the ureter are deposited on the surface of the DJ stents. Fragments deposited in layers lead to the closure of the catheter lumen, which impairs the process of the urine outflow [5].

Urinary tract infections are closely related to the process of adhesion. Binding bacteria to a ureteral stent is the main mechanism of biofilm formation. Aggregation of biofilm produced by the bacteria and precipitated urinary components causes kidney stones to form on the surface of ureteral stents [6].

The formation of deposits can lead to the occlusion of the catheter lumen, blocking the normal flow of urine. Furthermore, integration of stones covering the outer surface can adversely affect the removal procedure. According to the previous research, the layer formed on the surface of the catheter increases with longer implantation [5].

The construction of DJ stents is closely related to their functioning mechanism and their mechanical properties. Geometry is conformed to the urine flow mechanics, consisting of in-stent (luminal) and out-of-stent (extra luminal) flows.

Our previous research [5] investigated which part of a double-J ureteral stent (DJ stents) is most susceptible to deposition of post-URSL fragments and urea salt. Results showed that the proximal (renal pelvis) and distal (urinary bladder) part is the most amenable to post-URSL fragments and urea salt deposition. In addition, the proximal (renal pelvis) and distal (urinary bladder) parts present a higher tendency to encrustation than the middle part, thus that the highest stresses occur in these sections. A similar observation based on the flow catheter model [7] was confirmed by the occurrence of maximum stress located in the distal and proximal part of the DJ stents.

An important role in urinary outflow has side holes. Based on the flow catheter models, it was observed that the maximum stresses are concentrated around the holes [8]. Increasing the number of side holes increased the overall flow rate [5]. Stent thickness, side-hole shape and hole vertex angle also influence the encrustation level [8].

The diameter of the catheter is crucial in maintaining adequate urine flow after surgery. The selection of too small diameter creates the risk of stones clog. Furthermore, selecting the correct diameter ensures adequate urine flow pressure [9, 10]. DJ stents diameter is measured on the French (Fr) scale. One Fr equals 0.33 mm. Too low pressure will not push crushed stones into the bladder, while the overpressure can damage the catheter. The most commonly used diameters of DJ stents in clinical cases range from 5 Fr (1.66 mm) to 8 Fr (2.66 mm) [9]. Ureteral stents with a lower diameter such as 3 Fr (1.0 mm) and 4 Fr (1.33 mm) are used with children patients because of the smaller measure of the ureter in comparison to adult ureters [11].

The type of material used for fabricating ureteral stents significantly affects the process of adhesion [12]. The materials should comply with a number of requirements like smooth surface [13], high mechanical strength, should be flexible, biocompatible, non-toxic, have antibacterial properties, and prevent the adhesion of microorganisms and subsequent formation of biofilm, which promotes the formation of stones [14]. An important parameter of DJ catheters is their surface roughness. The higher the roughness, the more uneven the surface micropores, irregularities, and thickened areas can provide potential locations for bacteria binding. The most common materials used as ureteral stents are polyethylene, polyurethane and silicone [15]. The recommended choice of the material in this group is polyurethane [16].

In addition, to reduce encrustation, ureteral stents are coated with different substances like silver and gold nanoparticles thanks to their antibacterial properties [17]. Understanding the effect of encrustation requires a comprehensive approach. Coating the catheter surface with layers of stones causes more problems than just restricting the flow of urine. Discussed adhesion of bacteria and crushed stones also affect the mechanical properties of the ureter stent material. Furthermore, the surface roughness is associated with bacteria and crushed stone binding.

The novelty of this work was to determine the adsorption properties of the material used as urological stents, the cation affinity as a factor affecting encrustation, biofilm formation and to determine the effect of diameter on the accumulation of stone structures and crystals.

2 Materials and Methods

2.1 Materials

The tested subjects were 15 DJ urological stents placed during 15 procedures in the children patients after URSL therapy. DJ stents used in this research were made of polyurethane. The length of the examined catheters ranged from 8 to 32 cm (mean length 20 $\pm$ 2 cm). Double-J ureteral stents tested in this research were divided into three groups according to the value of the outer diameter: 1.00 (3 Fr), 1.33 (4 Fr) and 1.66 mm (5 Fr). The implantation period was 31 days.

2.2 Surface Analysis of DJ Stents

The microscopic analysis was performed on the stents before and after implantation (31 days). Energy-dispersive X-ray spectroscopy (EDS-energy dispersive X-ray spectroscopy) enables the analysis of the elemental composition of every sample.

The microscopic analysis was performed by the scanning electron microscope (SEM) JEOL JSM 7600F equipped with an X-ray analyzer INCA OXFORD. Before SEM observation, ureteral stents required an additional sample preparation procedure [12]. To improve sample conductivity the ureteral stent surface was covered with a 5 mm thick chromium layer.

To investigate the roughness on a smaller length scales a FlexAFM atomic force microscope with an Easyscan 2 controller (Nanosurf, Switzerland) was used. AFM (atomic force microscopy) measurements were made with noncontact mode, using the SICONA-10 cantrivalers (AppNano, USA). The measurements were illustrated by the use of the NanosurfEasyscan 2 software package. The study was carried out with three different samples. The same procedure as mentioned under Sect. 2.2 was followed with five randomly selected locations on the surface of the fixation plates for the scan area of 10 μm × 10 μm, with a resolution of 512 × 512 points.

2.3 Determination of Mechanical Properties

Mechanical properties such as tensile strength was tested for new stents and stents after implantation. The tensile strength was tested using an Zwick/Rockwell Testing System, load cell of 5 kN. DJ stents were tested with applied uniaxial tension with a testing rate of 1 mm/s for 1 s. A preconditioning run was done for each stent, including a 3-min hold time at 5 mm with 30 s between the preconditioning run and the first trial. DJ stents were studied using the tensile test and mounted the same way.

3 Results and Discussion

3.1 Influence of the Encrustation on DJ Stents Surface and Mechanical Properties

The microscopic observation (SEM/EDS), roughness measurements (AFM) and mechanical testing were performed on brand new DJ stents and after implantation in patients' bodies for 31 days. The surfaces of DJ stents made of polyurethane were tested using scanning electron microscopy and presented in Fig. 1, 2 and 3. The surfaces were examined before (Fig. 1A, B) and after implantation (Fig. 1D, E) for 31 days. Comparison of new ureteral stents and stents after implantation shows that the new stents have an uneven surface (Fig. 1C) that may enhance bacterial adhesion to their surface. Both the outer and inner surfaces of the stents are affected by the encrustation. To determine the surface roughness of urological catheters, they were examined with AFM. The average roughness parameters are presented in Table 1.

To define the effect of implantation on surface roughness, two parameters Sa (roughness average) and Sq (root mean square) were considered (Table 1B). The Sa parameter increased from value 2.06 nm to 2.72 nm. Sq parameter also increased from value 3.07 nm to 4.01 nm.

Despite the increase in values for Sa and Sq presented in Table 1B, the data obtained to state that the surface remains even (Fig. 1F). This appearance shows that crushed stones and crystals cover the surface of the entire stent equally. This proclaims that the process for stones and crystal's adherence to the surface of the stents is not only

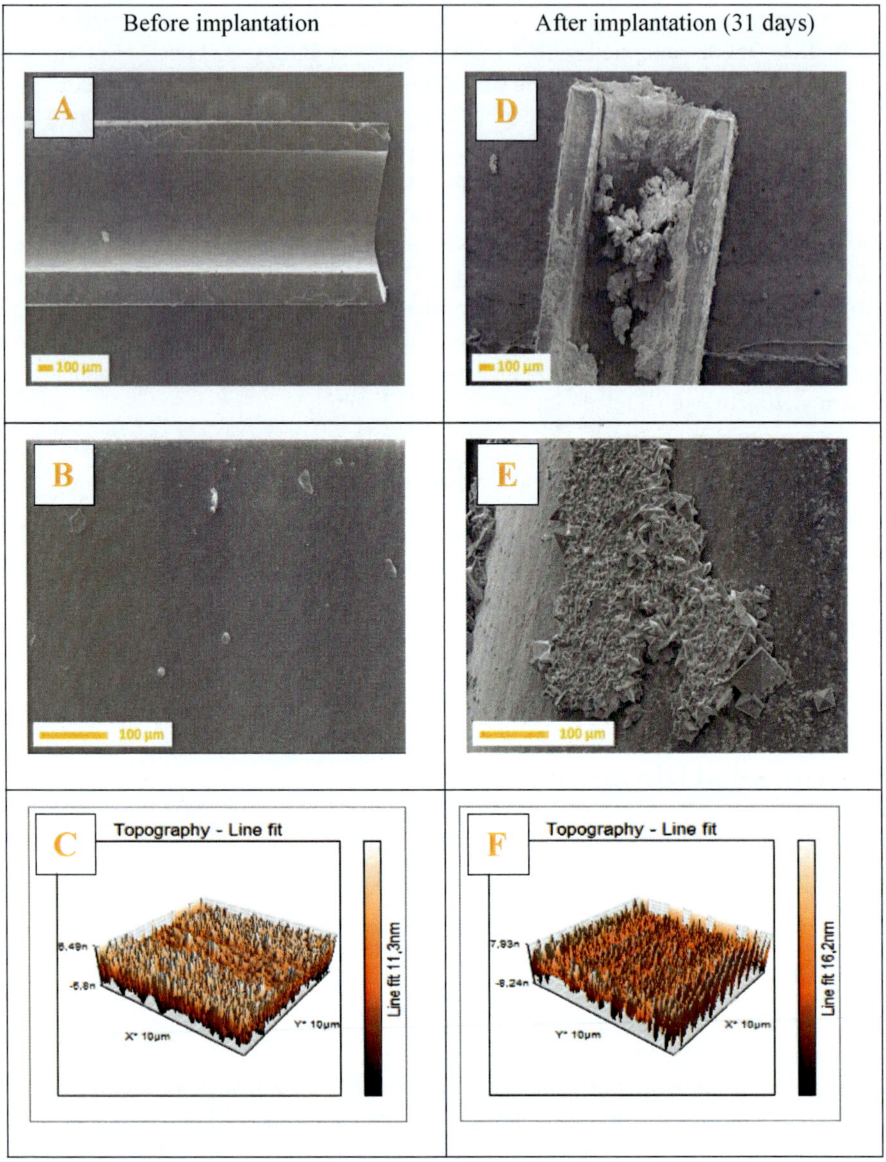

Fig. 1. SEM (A, B, D, E) and AFM (C, F) images of polyurethane DJ stents before and after implantation during 31 days.

determined by perfunctory/superficial stent's material construction flaws. The material from which the stents are made of – polyurethane exhibits the affinity to creating the biofilm and precipitation of crushed stones and crystals pieces on its surface. It confirms the finding of the previously conducted studies comparing two materials used for ureteral stenting.

Table 1. The results of tensile tests (Young's modulus and the breaking force), EDS and AFM analysis of DJ ureteral stent made of polyurethane, before and after implantation.

			Before implantation	After implantation
A	EDS	C [% wt.]	33.02±2.61	28.18±3.13
		N	17.00±1.80	18.43±3.87
		O	49.98±2.43	34.82±4.18
		P	---	1.96±0.77
		Ca	---	3.92±1.87
		Na	---	6.32±4.70
		Cl	---	6.37±3.82
B	Roughness	Sa [nm]	2.06±0.14	2.72±0.39
		Sq [nm]	3.07±0.61	4.01±0.47
C	Tensile tests	Young's modulus [kPa]	620±11.54	320±7.08
		Force [N]	18±0.57	7.5±0.09

Based on EDS results (Table 1A), it can be concluded that the main composition of urological stent is carbon, oxygen, nitrogen. Results of the EDS test (Table 1A) also showed that the inner and outer surfaces of the ureteral catheter were covered not only with the post-URSL fragments but also with the crystals of urea. According to an article by Dulawa [2], 40% of the stone deposits found in the urinary tract are calcium oxalates which confirms the EDS results obtained. Structures blocking the stents lumen are formed by the aftermath of urinary stone disruption with URSL, and the formation of small crystals growing on one another.

Moreover, a multi-layered structure built on the stent surface influences the stent's mechanical strength. Deterioration of the mechanical properties of ureteral stents can lead to fracture and breakage, which can complicate the removal process. The result of tensile tests of new stents and stents retrieved from patients after 31 days of implantation are presented in Table 1C. Young's modulus (E) was calculated for each DJ stent using engineering stresses. No change in cross-sectional area was assumed. After urological stent implantation, the value of Young's modulus and the force required for rupture is reduced by half.

The effect of implantation on the mechanical properties of DJ catheters was also studied by Gorman et al. [18]. They studied DJ stents made of polyurethane in the artificial fluid. According to article results, implantation significantly impacts on mechanical properties of DJ stents made of polyurethane. Furthermore, it also influences the decrease of Young's modulus value. Incrustation, which is the adhesion of crushed stone fragments and urine crystals to the surface of urological catheters, significantly impairs the mechanical properties of catheters made of polyurethane.

3.2 The Affinity of DJ Stents Part and Diameter on Encrustation

The polyurethane DJ stents implanted and retained for 31 days have been divided into three different sections: proximal part of the double-J ureteral stents situated in the renal pelvis, middle part, and distal part situated in the urinary bladder. Each DJ stent was covered with a multilayered compact structure of crushed stones (both on inside and outside surface), as shown in Figs. 2, 3 and 4. The most extensive agglomerations were observed in Figs. 2A, 3A and 4A showed ureteral stent after 31 days of implantation with a diameter of 1.0 mm (3 Fr). Ureteral stents with a diameter of 1.33 mm (4 Fr) were also covered by small calculi growing on one another (Figs. 2B, 3B and 4B). The 1.66 mm diameter (5 Fr) stents were covered with significantly reduced stone structures (Figs. 2C, 3C and 4C) compared to the 1.00 mm and 1.33 mm diameter stents.

As can be seen in Figs. 2, 3 and 4 the inner surface of ureteral stents is covered with more complex biofilm and crushed stones after the URSL procedure. The biggest incrustation was observed from the analysis of 3 different diameters for 1.0 mm diameter (Figs. 2A, 3A and 4A). The lowest accumulation of crushed stones and crystals on the catheter surface was observed for a diameter of 1.66 mm (Figs. 2C, 3C and 4C), therefore this stents were taken into account during later analyses.

Results developed by Kim K. et. al. indicating that DJ stents with bigger diameter provide better urine flow [9] confirm presented results. A better urine flow with crushed stones is associated with reducing adhesion of crushed stones after URSL procedure which can lead to encrustation.

DJ stents with diameter of 1.66 mm (Figs. 2C, 3C and 4C) were analysed with separation for 3 parts mentioned above. Analysed SEM results showed that the encrustation at the proximal end is bigger than at the distal coil (Fig. 4C). The proximal part (Fig. 2C) of the stent is located in patient renal pelvis. Developed results are confirmed by Kim et al. [7] who were testing the urine flow with different side holes aggregation. According to their research the last side hole in the proximal coil of the stent and the even side hole throughout the ureter are critical to a DJ stent to maximize its role in the upper urinary system. Similar observation was made in the research made by Sighinolfi et al. [19]. Moreover the distal (Fig. 4C) part is located at the very beginning of the crushed stone path and is therefore the first agglomeration site after surgery.

SEM analysis (Figs. 1, 2, 3 and 4) graphically documented the changes on the DJ stents surface after implantation for 31 days. The outer and inner surfaces of the DJ stents are covered with multilayer structures. Additionally, EDS analysis (Table 1A) confirmed the assumptions that the structures present on the surface of the stents are oxalate stones. Moreover inner and outer surfaces of DJ stents were covered with the crystals of urea and by the post-URSL fragments.

AFM results (Table 1B) showed that after implantation surface of DJ stents is higher roughness. Similar relationship/dependence was observed by Palka et al. [20] in their work studying the stents after the implantation. Higher roughness surface can increase adhesion of uropathogens causing urinary infection and adsorption of crushed stones after URSL procedure.

Studying the results of the tensile tests of the stents has shown (Table 1C) an impairment of implantation on the values of Young's modulus and the force leading

Diameter [mm]		Proximal outer part	Proximal inner part
A	1.00		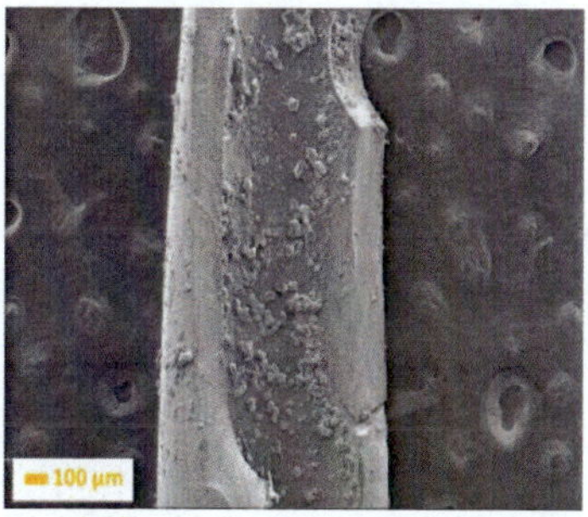
B	1.33	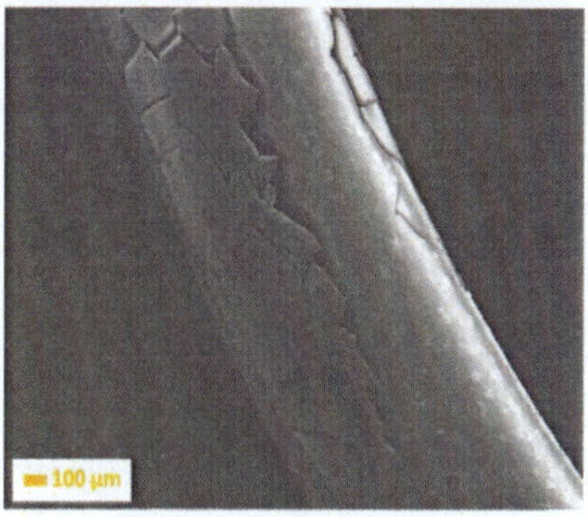	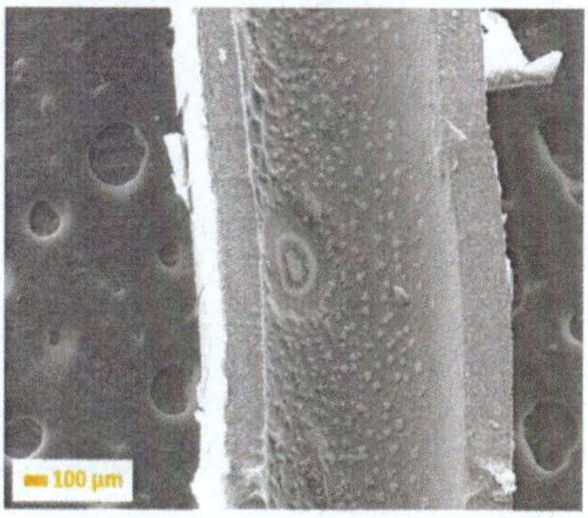
C	1.66		

Fig. 2. SEM images of a polyurethane DJ stents with different diameters (proximal part) after the implantation of 31 days.

to stent breakage. Obtained results are consistent with the literature [18, 21, 22]. Therefore, implantation can lead to DJ stents breakage, which can complicate the process of removing the stents and cause patients pain.

According to the author's knowledge, there are no scientific reports on the correlation between the stone layer composition and surface roughness of DJ stents. The key consideration found was the roughness associated with the adhesion of uropathogens to the stent surface [16]. Based on the research results it can be concluded that polyurethane is

Diameter [mm]		Middle outer part	Middle inner part
A	1.00		
B	1.33		
C	1.66		

Fig. 3. SEM images of a polyurethane DJ stents with different diameters (middle part) after the implantation of 31 days.

not the appropriate material for ureteral stents, what was confirmed by our latest research [5].

The comparison of DJ stents having different diameter has been confirmed that the encrustation increases with lower diameter of the ureteral stent. Less accumulation of biofilm, stone structures and precipitated crystals from the urine was observed for the 5 Fr (1.66 mm) DJ stents, which also indicates that its flow was the most efficient. That result

Diameter [mm]		Distal outer part	Distal inner part
A	1.00		
B	1.33	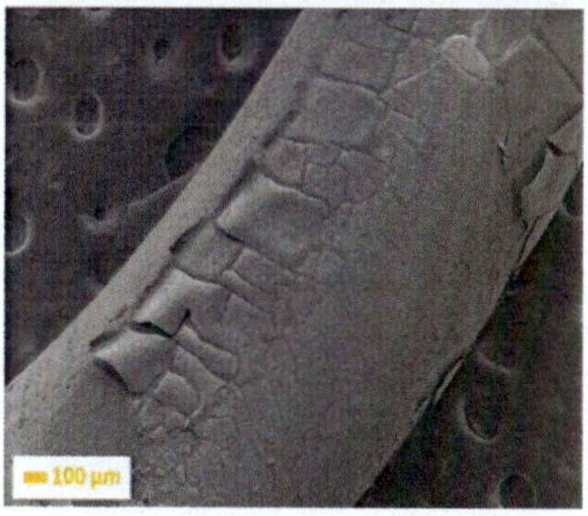	
C	1.66	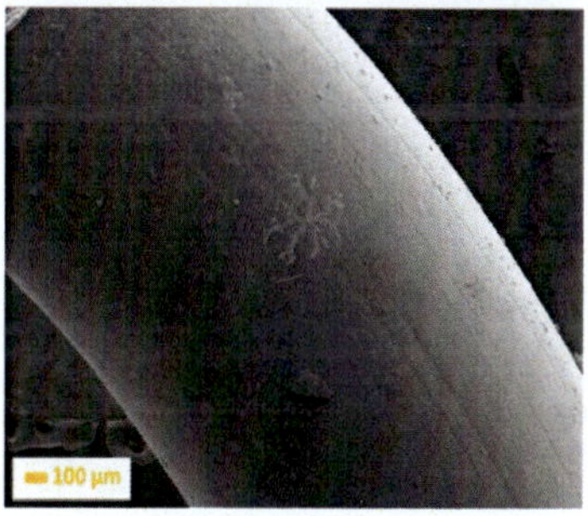	

Fig. 4. SEM images of a polyurethane DJ stents with different diameters (distal part) after the implantation of 31 days.

is confirmed by work of Hyoung-Ho Kim et al. [23]. They investigated which sizes of double-J stents are more effective in achieving an acceptable urine flow.

The available articles look at particular aspects of encrustation. As was mentioned earlier in this work, understanding encrustation of DJ stents needs comprehensive approach. Choice material that comply with requirements mentioned under Sect. 1 is not enough. Selecting the appropriate DJ stent diameter is a value closely related to urine flow quality. In addition, to reduce encrustation, ureteral stents need coated with different substances such as silver and gold nanoparticles because of their antibacterial properties [17, 24, 25], metallic nanoparticles [25], nitrofurazone [26], chlorhexidine [27], polytetrafluoroethylene (PTFE), hydrogel [13] and lysostaphin [28].

4 Conclusions

The aim of this work was to determine the adsorption properties of the polyurethane double-J stent and determine the effect of stent's diameter on the accumulation of stone structures and crystals. Encrustation depends on the diameter of DJ stents – lower DJ stents diameter increases risk of their blockage.

The implantation through the period of 31days affected tensile tests results, after implantation, the value of Young's modulus was reduced from 620 kPa to 320 kPa, and the force required for rupture is reduced from a value of 18 N to 7.5 N. Furthermore, implantation affects the surface by increasing its roughness. DJ stent diameter (1.00, 1.33, 1.66 mm) impacts the encrustation phenomenon: the biggest aggregation of crushed stones and crystals was observed for the DJ stents with the diameter of 1.00 mm, and the least encrustation was observed for 1.66 mm. EDS results confirm the presence of calcium oxalates stones on the surface of DJ stents.

Performed results indicate the need for stents surface coating with substances providing antibacterial properties and lower affinity to adhesion crushed stones and crystals.

References

1. Halinski, A., et al.: Interest of URS-L in the treatment of ureterolithiasis in preschool children. Front. Pediatr (2019). https://doi.org/10.3389/fped.2019.00324
2. Duława, J.: Czynniki rozwoju kamicy nerkowej. Forum Nefrologiczne **2**(3), 184–188 (2009)
3. Bochniewska, V., Jung, A., Jurkiewicz, B., Jobs, K.: Metody zabiegowe w leczeniu kamicy układu moczowego u dzieci. Pediatr. Med. Rodz. **6**(4), 309–314 (2010)
4. Kupajski, M., Tkocz, M., Ziaja, D.: Modern management of stone disease in patients with a solitary kidney. Videosurg. Other Miniinvas. Tech. **1**, 1–7 (2012). https://doi.org/10.5114/wiitm.2011.25641
5. Arkusz, K., Pasik, K., Halinski, A., Halinski, A.: Surface analysis of ureteral stent before and after implantation in the bodies of child patients. Urolithiasis **49**(1), 83–92 (2020). https://doi.org/10.1007/s00240-020-01211-9
6. Arkusz, K., Krasicka-Cydzik, E.: The effect of phosphates and fluorides, included in TiO_2 nanotube layers, on the performance of hydrogen peroxide detection. Arch. Metall. Mater. **63**(2), 765–772 (2018)
7. Kim, K.-W., Kim, H.-H., Choi, Y.H., Lee, S.B., Baba, Y., Suh, S.-H.: Arrangement of side holes in a double J stent for high urine flow in a stented ureter. J. Mech. Sci. Technol. **34**(2), 949–954 (2020). https://doi.org/10.1007/s12206-020-0144-1
8. Mosayyebi, A., et al.: Reducing deposition of encrustation in ureteric stents by changing the stent architecture: a microfluidic-based investigation. Biomicrofluidics **13**(1), 014101 (2019). https://doi.org/10.1063/1.5059370

9. Kim, K., Kim, H., Choi, Y.H., Lee, S.B., Baba, Y.: Urine flow analysis using double J stents of various sizes in in vitro ureter models. Int. J. Num. Methods Biomed. Eng. **36**(2), 1–12 (2020). https://doi.org/10.1002/cnm.3294
10. Nestler, S., Witte, B., Schilchegger, L., Jones, J.: Size does matter: ureteral stents with a smaller diameter show advantages regarding urinary symptoms, pain levels and general health. World J. Urol. **38**(4), 1059–1063 (2019). https://doi.org/10.1007/s00345-019-02829-0
11. Raisin, G., et al.: Does a small stent size increase the risk of post-operative complications following reconstructive robotic surgery in the pediatric population? J. Pediatr. Endosc. Surg. **2**(1), 11–14 (2020). https://doi.org/10.1007/s42804-020-00040-2
12. Nycz, M., Paradowska, E., Arkusz, K., Kudlinski, B., Krasicka-Cydzik, E.: Surface analysis of long-term hemodialysis catheters made of carbothane (poly(carbonate)urethane) before and after implantation in the patients' bodies. Acta Bioeng. Biomech. **20**(2), 47–53 (2018)
13. Lawrence, E.L., Turner, I.G.: Materials for urinary catheters: a review of their history and development in the UK. Med. Eng. Phys. **27**(6), 443–453 (2005). https://doi.org/10.1016/j.medengphy.2004.12.013
14. Tunney, M., Gorman, S.: Evaluation of a poly(vinyl pyrollidone)-coated biomaterial for urological use. Biomaterials **23**(23), 4601–4608 (2020). https://doi.org/10.1016/s0142-9612(02)00206-5
15. Scarneciu, I., et al.: The risk factors and chemical composition of encrustation of ureteral double j stents in patients with urolithiasis. Rev. Chim. **69**(12), 3406–3409 (2018). https://doi.org/10.37358/RC.18.12.6759
16. Venkatesan, N., Shroff, S., Jeyachandran, K., Doble, M.: Effect of uropathogens on in vitro encrustation of polyurethane double J ureteral stents. Urol. Res. **39**(1), 29–37 (2010). https://doi.org/10.1007/s00240-010-0280-7
17. Arkusz, K., Nycz, M., Paradowska, E., Pijanowska, D.G.: Electrochemical stability of TiO_2 nanotubes deposited with silver and gold nanoparticles in aqueous environment. Environ. Nanotechnol. Monitor. Manag. **15**, 1–12 (2021). ISSN: 2215-1532
18. Gorman, S.P., Jones, D.S., Bonner, M.C., Akay, M., Keane, P.F.: Mechanical performance of polyurethane ureteral stents in vitro and ex vivo. Biomaterials **18**(20), 1379–1383 (1997). https://doi.org/10.1016/s0142-9612(97)00070-7
19. Sighinolfi, M.C., et al.: Chemical and mineralogical analysis of ureteral stent encrustation and associated risk factors. Urology **86**(4), 703–706 (2015). https://doi.org/10.1016/j.urology.2015.05.015
20. Palka, L., Mazurek-Popczyk, J., Arkusz, K., Baldy-Chudzik, K.: Susceptibility to biofilm formation on 3D-printed titanium fixation plates used in the mandible: a preliminary study. J. Oral Microbiol. **12**(1), 1838164 (2020). https://doi.org/10.1080/20002297.2020.1838164
21. Hendlin, K., Dockendorf, K., Horn, C., Pshon, N., Lund, B., Monga, M.: Ureteral stents: coil strength and durometer. Urology **68**(1), 42–45 (2006). https://doi.org/10.1016/j.urology.2006.01.062
22. Pedro, R.N., Hendlin, K., Kriedberg, C., Monga, M.: Wirebased ureteral stents: impact on tensile strength and compression. Urology **70**(6), 1057–1059 (2007). https://doi.org/10.1016/j.urology.2007.08.007
23. Kim, H.H., Kim, K., Choi, Y.H., Lee, S.B., Baba, Y.: Numerical analysis of urine flow with multiple sizes of double-J stents. Appl. Sci. **10**(12), 4291 (2020). https://doi.org/10.3390/app10124291
24. Arkusz, K., Paradowska, E., Nycz, M., Mazurek-Popczyk, J., Baldy-Chudzik, K.: Evaluation of the antibacterial activity of Ag- and Au-nanoparticles loaded TiO_2 nanotubes. J. Biomed. Nanotechnol. **16**(9), 1416–1425 (2020)
25. Nycz, M., Arkusz, K., Pijanowska, D.G.: Influence of the silver nanoparticles (AgNPs) formation conditions onto titanium dioxide (TiO_2) nanotubes based electrodes on their impedimetric response. Nanomaterials **9**(8), 1072 (2019). ISSN: 2079-4991

26. Lo, J., Lange, D., Chew, B.: Ureteral stents and foley catheters-associated urinary tract infections: the role of coatings and materials in infection prevention. Antibiotics **3**(1), 87–97 (2014). https://doi.org/10.3390/antibiotics3010087
27. Zelichenko, G., et al.: Prevention of initial bioflm formation on ureteral stents using a sustained releasing varnish containing chlorhexidine. Vitro study. J. Endourol. **27**(3), 333–337 (2013). https://doi.org/10.1089/end.2012.0193
28. Kotaskova, I., et al.: Molecular techniques complement culture-based assessment of bacteria composition in mixed biofilms of urinary tract catheter-related samples. Front. Microbiol. (2019). https://doi.org/10.3389/fmicb.2019.00462

Standardisation Procedure of Infra-red Imaging in Biomechanics

Bartłomiej Zagrodny(✉)

Lodz University of Technology, ul. Stefanowskiego 1/15, 90-924 Łódź, Poland
bartlomiej.zagrodny@p.lodz.pl

Abstract. Purpose: The methods of thermal imaging in biomechanics and medicine are still not fully standardised, in contrary to other popular methods of muscle activity examination. This this makes it difficult to compare test results performed in different laboratories or even undermines credibility of some of the results.

Methods: The proposed standardisation procedure is based on the International Association of Certified Thermographers, American Academy of Thermology, European Association of Thermology, scientific publications and authors experience. The most restricted recommendations are chosen, described and discussed.

Results: The standardisation procedure of infra-red imaging in biomechanics is presented and discussed. Volunteer preparation, laboratory conditions, and tips how to make a thermal image are described. The state of the art is presented and chosen aspects of thermal imaging pointed out.

Conclusions: The presented method allows to create repeatable conditions and to minimise the influence of factors that can change the temperature readings. As a result, when using the proposed standardisation, higher repeatability and precision of measurements are obtained.

Keywords: Thermal imaging · Biomechanics · Musculo-skeletal

1 Introduction

Modern thermography is a relatively young science and it can be stated, that it starts in the early 30s of the 20th century, despite the infra-red radiation is a much earlier discovery. During the last 100 years, the thermal imaging technique gains in popularity and is often used for many purposes like security, control of electrical systems, buildings insulations, finding wet/humid areas. Nevertheless, its most prominent use should be identified with medicine and biomechanics. Detecting symptoms of diseases, vascular problems, or even scanning for cancer is one of the most common use. In biomechanics scanning muscle systems for its activity, overloading, faulty posture or musculoskeletal injury is just one of the most common applications [1, 2]. Definitely, this technique is getting more and more popular in the field of biomechanics; one can find plenty of research on this topic. According to Google Scholar, there is about 18,000 answers for the question "thermal imaging muscle system".

A. Hadamus et al. (Eds.): BIOMECHANICS 2021, LNNS 328, pp. 145–153, 2022.
https://doi.org/10.1007/978-3-030-86297-8_13

The biggest advantage of infra-red is that it is a non-contact technique. Unfortunately, when speaking about a musculoskeletal system (human or animal), researchers can expect many variables that can influence the experiment results. When analysing State of the Art of thermal imaging regarding humans, one can find some propositions of standardization procedures. International Association of Certified Thermographers, American Academy of Thermology, European Association of Thermology, and papers [3, 4] are just examples of them. Nevertheless, it is worth to notice, that in recently published and analysed papers (see for example [3–5, 11, 12] authors are using different methods of volunteer preparation and providing conditions in laboratories, which sometimes even they are changing during an experiment. This generates a question if the published results are repeatable or even reliable. According to the author's knowledge, the most advanced (and restricted) procedure is presented in [4], nevertheless, it can be improved and supplemented with new findings.

The aim of this paper is to present a method of infra-red experiments standardization in biomechanics. Method of volunteer and laboratory facility preparation along with guidelines for making thermal images are presented and described.

2 Standardisation Procedure of Thermal Imaging

The main points of the standardisation procedure are based on the International Association of Certified Thermographers, American Academy of Thermology, European Association of Thermology, and papers [4–6]. The most restricted advices are chosen and supplemented with recommendations created based on the author own experience.

2.1 Volunteer Preparation

The most important element of the standardization procedure is the correct preparation of the volunteer. The main recommendations are pointed out in Table 1. Listed recommendations should help to minimize the influence of external factors or the behaviour of the volunteer on the skin temperature prior to the experiment. The most important ones are connected with the actions which effects persist for a long time, especially like sunburns or muscle recovery after an extensive workout. Some of the recommendations, like not to use any lotions, creams, detergents, or similar minimum 3 days before the experiment should help to stabilize emissivity for body areas on the same level [7]. If the thermal image is to be done on a hairy area, hair should be removed three days before, preferably by shortening them at skin level, but not shaving, which can cause local inflammation. About one day before the experiment, the volunteer should stop consuming alcohol and should start to avoid greasy meals or other meals, spices, and substances that lead to greater thermogenesis (ex. spices with capsaicin, black pepper, ginger, mixed spices, and drinks like green or black tea and caffeine) [8]. Tight garment parts can restrict blood flow [13] and as a result change body segment temperature, what is the reason why they should be avoided before thermal imaging experiments.

Additionally, volunteer's skin should be free from any inflammations, wounds, cuts, and other types of skin damage. Some tattoos can modify temperature readings, which makes it necessary for screening tattooed areas for unexpected, local differences in temperature [14]. Core temperature should not exceed 37 °C. It is recommended to have an interview regarding all volunteer injuries. Some of them, even considered as insignificant can have long-lasting (even life-lasting) effect, however, finding volunteer without any injury-history can be a real challenge, thus the researchers should be aware of a possible influence of the injury and faulty posture history on the thermal imaging.

Table 1. Main recommendation of the volunteer preparation.

Time before experiment (minimum)	Action
5 days	Do not sunbath, do not use sauna or cryochamber Stop using detergents when taking shower Avoid hot or cold baths
3 days (72 h)	Do not use: Creams, lotions, powders, perfumes, deodorants, antiperspirants and other similar substances Avoid: Physical exercise, intensive physical activity, massage, electrical muscles/nerves stimulation, ultrasound examination, acupuncture, use of warm or cold compresses etc Remove hair from the study area
1 day (24 h)	Do not consume alcohol, Avoid greasy, spicy meals and or other substances that can improve thermogenesis Do not use tight clothing items
12 h	Last bath: a short shower in lukewarm water without detergents
4 h	Avoid physical exertion (quick gait, run, stairs climbing to higher floors, exercises) Do not consume drinks with caffeine (including strong tea), hot or cold drinks Last meal, avoid high caloric one
2 h	Do not eat, but avoid experiment on empty stomach

2.2 Laboratory Conditions

Not only the volunteer has to be prepared for the experiment. Very important is to correctly prepare the laboratory facility. The key actions are presented in Table 2. At first, the temperature and real humidity should be stabilised. It is advised to keep the room temperature in the range 21 °C–24 °C [9] and to allow the volunteer to choose his/her preferred temperature during the experiment in this range to ensure a proper thermal comfort level. The temperature should not change during the experiment. Temperature below 18 °C can cause shivering, above 24 °C – extensive sweating. Highly efficient heat sources, ACs, fans, humidifiers, or air dryers may be used, but they should be removed or turned off during the experiment. Especially IR radiators, like heaters should be removed or covered. On the day of the experiment, the advection and convection in the laboratory should be minimised to avoid cooling air movement. It is also advised to use window blinds, the best is with IR-mirrors. If it is possible, the best placement for the laboratory, where this type of experiment is performed is a northern site of the building or room without windows.

Table 2. Main recommendations for the laboratory preparation.

Time before experiment (minimum)	Action
2 days (48 h)	Stabilise: Temperature in range 21 °C–24 °C Real humidity in range 40–55%
1 day (24 h)	Remove all IR raditors (if their temperature is higher than 10 °C than temperature in the laboratory) Cover unmovable IR radiators, especially working heaters
In a day of experiment	Reduce to the possible minimum: Convection Advection Use window blinds in the lab to reduce the external IR radiation Do not use ventilators, AC or similar equipment which can change the temperature or RH rapidly or causes an intensive air movement

2.3 Making the Thermal Image

Last but not least is to make a proper thermal image. Main recommendations are presented in Table 3. After the volunteer arrives at the laboratory an adaptation period should be performed. Volunteers stay in a garment in which thermal image will be done. No additional covers of the skin or body parts are allowed. The volunteer should not touch or leaning anything that can change his/her skin/body temperature. The position should be natural and symmetrical. No muscle part should be overload ex. by standing or sitting in an inconvenient or asymmetrical position. A slow gait is allowed.

The minimal time of adaptation is 15 min and it can be calculated according to the equation:

$$\tau_{min}[\text{min}] = 15 + (|T_{ext} - T_{int}|)$$

where: τ_{min} is minimal time of adaptation in minutes, T_{ext} and T_{int} – external (outside laboratory) and internal (in the laboratory) temperature in Celsius/Kelvin degree respectively. For the Fahrenheit scale, this rule should be adequately modified. The camera should be placed on stable support at a constant and repeatable distance from the volunteer. Its optical axis should be placed with respect to the normal of the object surface (ex. torso) due to the Lambertian nature of IR radiation [9]. The emissivity should be chosen for proper skin pigmentation. The average and accepted values in the literature are in range 0.97–0.98 [10].

It is also advised to choose a neutral background, for example, a wall with uniform temperature. Moreover, a gold standard should be to check and store all necessary data like temperature and humidity in the laboratory, date, and time of the experiment. A good practice should be to check the body fat of the volunteer (amount and distribution) and to make basic anthropometric measurements. Taking photos in S and F plane should facilitate the analysis. A questionnaire about performed sports/activities should also help to analyse and understand results.

Table 3. Main recommendations for making thermal image.

Action	Advise
Thermal camera placement	Use a tripod Make thermal image with use of a uniform background
Volunteer placement in a front of camera	Keep constant distance from the volunteer for all captured thermal images Place camera optical axis with respect to the normal of the object surface Choose a proper duration of thermal adaptation
Setting the emissivity	Constant for all recordings at the level of 0.97–0.98

2.4 Control List

It is advised to create a control list. It should help to understand all problems found during the analysis of thermal images and to distinguish between technical problems/errors in procedure and natural phenomena. An example of this type of control list, created on the basis of control list (translated and rearranged) used in the laboratory of Biomechanics Lodz University of Technology is presented in Supplementary Materials. It includes the most important questions about volunteer activity, faulty posture, photography in anatomical position (Table S1) and checklist (Table S2).

3 Conclusions

Standardisation of examination procedures is crucial for obtaining reliable and repeatable results. For some of the techniques, conditions are restricted. For example, when sEMG is made not according to the SENIAM regulations this makes results unpublishable. The variety of methods, equipment, conditions in the laboratory, and procedures in different research centres makes published results of thermal analysis often incomparable and even questionable. Such a standard procedure in this field is mandatory. Presented procedure should help to obtain a repeatable conditions of thermal imaging experiments and to obtain more reliable results. Main recommendations are presented in Tables 1–3.

Supplementary Materials

Example of the checklist and questionnaire

Date:________

№ of volunteer (code):____________________
Name/surname: (if the code system is not used)
:____________________________

Hour: arrival of the volunteer:__________/departure: __________

Temperatures:
in the lab.: beginning of experiment__________, end of experiment_________
external: __________

Humidity in the lab.:
beginning of experiment__________, end of experiment_________

Time of adaptation: __________min.

Camera model used during experiment:__________
Emissivity:_________

Anthropometry:

Sex: _________, Age: _________, Height: ________cm, Mass: _________kg,

Body fat: __________%

Other notes:____
____
____
____
____
____
____
____

Table S1. Photography in anatomical position.

F plane, front of the body	F plane, back of the body	S plane, left site of the body	S plane, right site of the body

Sports and activities performed by volunteer:

____

____

____

Is any of the activity asymmetrical? YES / NO

Faulty postures:

____

____

____

____

Is leg length equal? YES / NO if no, then the L / R leg is longer by ________ cm.

Table S2: Preparation.

Time before experiment (minimum)	Action	Fulfilled Yes / No	Notes: Ex. antiperspirant was used
5 days	Do not sunbath, do not use sauna or cryochamber Stop using detergents when taking shower Avoid hot or cold baths	Yes / No Yes / No Yes / No	
3 days (72h)	Do not use: creams, lotions, powders, perfumes, deodorants, antiperspirants and other similar substances Avoid: physical exercise, intensive physical activity, massage, electrical muscles / nerves stimulation, ultrasound examination, acupuncture, use of warm or cold compresses, etc. Remove hair from the study area	Yes / No Yes / No Yes / No	
1 day (24h)	Do not consume alcohol, Avoid greasy, spicy meals and or other substances that can improve thermogenesis Do not use tight clothing items	Yes / No Yes / No Yes / No	
12h	Last bath: a short shower in lukewarm water without detergents	Yes / No	
4h	Avoid physical exertion (quick gait, run, stairs climbing to higher floors, exercises) Do not consume drinks with caffeine (including strong tea), hot or cold drinks Last meal, avoid high caloric one	Yes / No Yes / No Yes / No	
2h	Do not eat, but avoid experiment on empty stomach	Yes / No	

Other notes:____

Examiner:

References

1. Somboonkaew, A., et al.: Mobile-platform for automatic fever screening system based on infrared forehead temperature. In: Proc. Opto-Electron. Commun. Conf. (OECC) Photon. Global Conf. (PGC), pp. 1–4 (2017)
2. Vollmer, M., Möllmann, K.-P.: Infrared Thermal Imaging: Fundamentals, Research and Applications, 2nd edn. Wiley, USA (2017)
3. Bauer, J., Dereń, E.: Standardization of thermographic studies in medicine and physical therapy. Acta Bio-Optica Inform. Med. Biomed. Eng. **20**(1), 10–12 (2014). in Polish
4. Moreira, D.G., et al.: Thermographic imaging in sports and exercise medicine: a Delphi study and consensus statement on the measurement of human skin temperature. J. Thermal Biol. **69**, 155–162 (2017)
5. Coletta, N.A., et al.: Core and skin temperature influences on the surface electromyographic responses to an isometric force and position task. PLoS ONE **13**(3), e0195219 (2018)
6. Chudecka, M., Lubkowska, A.: Temperature changes of selected body's surfaces of handball players in the course of training estimated by thermovision, and the study of the impact of physiological and morphological factors on the skin temperature. J. Thermal Biol. **35**(8), 379–385 (2010)
7. Steketee, J.: The influence of cosmetics and ointments on the spectral emissivity of skin (skin temperature measurement). Phys. Med. Biol. **21**, 9–20 (1976)
8. Westerterp-Plantenga, M., et al.: Metabolic effects of spices, teas, and caffeine. Physiol. Behav. **89**(1), 85–91 (2006)
9. Priego Quesada, J.I., et al.: Effects of graduated compression stockings on skin temperature after running. J. Thermal Biol. **52**, 130–136 (2015)
10. Jones, B.F., Plassmann, P.: Digital infrared thermal imaging of human skin. IEEE Eng. Med. Biol. Mag. **21**, 41–48 (2002)
11. Zagrodny, B., et al.: Could thermal imaging supplement surface electromyography measurements for skeletal muscles? IEEE Trans. Instrum. Meas. **70**, 1–10 (2021)
12. Kuniszyk-Jóźkowiak, W., Jaszczuk, J., Czaplicki, A.: Changes in electromyographic signals and skin temperature during standardised effort in volleyball players. Acta Bioeng. Biomech. **20**(4), 115–122 (2018)
13. Xiong, Y., Tao, X.: Compression garments for medical therapy and sports. Polymers **10**, 663–668 (2018)
14. Zagrodny, B., Kaczorowski, Łukasz, Awrejcewicz, J.: Influence of Body Tattoo on Thermal Image—A Case Report. In: Gzik, M., Paszenda, Z., Pietka, E., Tkacz, E., Milewski, K. (eds.) AAB 2020. AISC, vol. 1223, pp. 209–214. Springer, Cham (2021). https://doi.org/10.1007/978-3-030-52180-6_23

Can Tattoo Influence a Thermal Image? A Case Report

Bartłomiej Zagrodny(✉)

Lodz University of Technology, ul. Stefanowskiego 1/15, 90-924 Łódź, Poland
bartlomiej.zagrodny@p.lodz.pl

Abstract. Background: The aim of this work is to answer the question if a body tattoo can influence thermographic examination and if the results for a new and old tattoo are different. This work is also a continuation of an earlier study devoted to this problem.

Methods: The study was conducted on one volunteer with a tattoo made in a professional tattoo parlour, on the upper part of his chest 16 months and 28 months after making a tattoo. To record a skin temperature distribution, an infrared camera was used.

Results: It is shown that a tattoo may have an influence on a local skin temperature distribution. The main applicable conclusion is that during any kind of thermographic examination, the examiner should expect that the tattoo can locally change temperature readings.

Keywords: Thermal imaging · Tattoo · Influence of body-tattoo on temperature distribution

1 Introduction

According to [1], it is estimated that around 25% of the world population has at once one tattoo. In Europe, the data vary from 5 up to 10% [2]. This type of body modification is also relatively popular in the United States. According to [9] from 21% up to 29% of the population of the United States have at least one tattoo and 15–20% have two or more. Tattoos are most common at a young age, which is consistent with the data published for European society. In Poland, it is estimated that the percentage of tattooed population age 15–50 is estimated at the level of 9%. Around 30% of people without a tattoo declare that they want to make at least one in the future, but this percentage of people who declare this decreases with age. Around 33% of people with a tattoo have one, 30% two tattoos, 20% three or more [3].

It is a well-known fact that tattoos can cause different types of dermatological issues, often caused by a tattoo ink [4, 5]. If the tattoo caused an inflammation or allergic reaction [10], then it is obvious that the examiner should expect a higher local temperature. However, the problem of existence of a tattoo during thermal imaging of the body/skin is still a relatively weakly examined area, often neglected during this type of experiments/measurements. According to the previous work [11], the tattoo can locally

A. Hadamus et al. (Eds.): BIOMECHANICS 2021, LNNS 328, pp. 154–158, 2022.
https://doi.org/10.1007/978-3-030-86297-8_14

influence the readings of an infrared camera, thus the existence of tattoos can-not be neglected. For this study, an infrared camera was used. This method of skin and body examination is used in medicine (ex. in dermatology [7]) and biomechanics (ex. [6, 8]), also for precise examination of muscle activity [12]. In that case, even a small temperature difference reading can be misleading. Another problem exists during the standardisation of the thermal examination procedure regarding tattoos. Only found in the literature authors of [6,7] advise to wait with the experiment until the skin healing process is finished after tattooing. At this moment, we know that the influences of the body tattoo on temperature distribution exist, but we do not know if this influence is changing in time.

The aim of this work is to check if a body tattoo can influence thermographic examination and if the results for a new and old tattoo are different.

2 Materials and Methods

The thermal image of the tattooed area (chest) was done the first time after 16 months (later 1. case) and the second time after 28 months (later 2. case) after tattooing. In both cases no dermatological problem was found. Also, in both cases, the same protocol, camera and technique were used to obtain thermal images. During the examination Nec-Avio R300SR-S with FPA-type sensor, spectral range 8–14 μm and NETD 0.08K were used. Thermal images were done from a distance of 0.1 m. The volunteer was prepared according to the protocol described in [12], which is based on the requirements of International Association of Certified Thermographers, American Academy of Thermology, European Association of Thermology, and [6], with additional modifications. The laboratory conditions were as follows:

i. Temperature in the laboratory 22.5 $\pm$ 0.1 °C;
ii. Humidity 45% $\pm$ 5 RH;
iii. Advection and convection reduced to minimum;

Data of the volunteer are as follows:

i. Age: 21 (first cycle of measurements)/22 (second cycle of measurements);
ii. Sex: male;
iii. Tattoo placement: upper part of the chest;
iv. Type of the tattoo: black ink inscription done in a professional Tattoo-parlour;
v. Body fat: 12% according to the measurement with 3-point skin fold calliper;
vi. Body core temperature: 36.6 °C;
vii. Body-hair removed 72 h before experiment.

In both cases a 20 min of adaptation period was applied.

3 Results

Figure 1 presents photography of the volunteer chest with tattoo inscription that undergoes further examination. Figure 2 presents tattoo superimposed on photography of the same area for the first measurements – 16 months after tattooing (left) and second measurements – 28 months after tattooing (right) for the same thermal camera settings and thermal images post processing method. Dimensions of the description are as follows: "R" height 60 mm, "a" height 40 mm. In both cases (1. and 2.) whole description was visible in infra-red, but as it was mentioned in [11] not all parts of the sentence and even parts of letters have the same temperature difference. It was observed that the areas with a higher amount of the ink had higher temperature differences. The volunteer has confirmed that most of these places were refilled with ink after 3–4 weeks from tattooing.

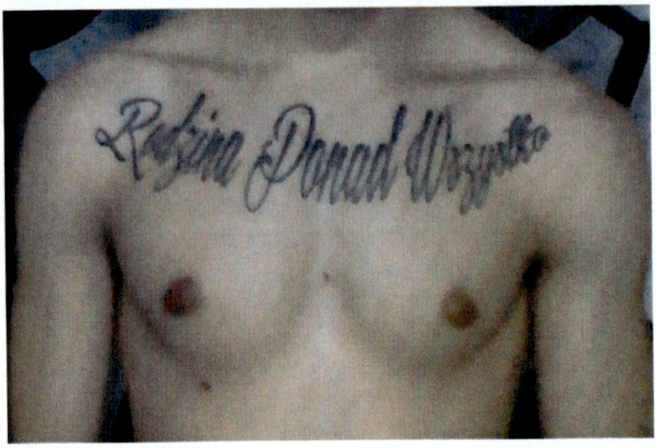

Fig. 1. Photography of the examined area.

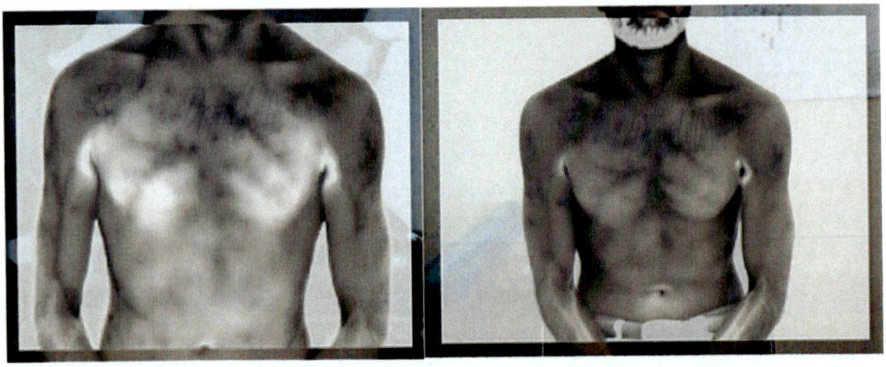

Fig. 2. Photography of the examined area with superimposed thermal image 16 months after tattooing (left) and second measurements – 28 months after tattooing (right).

Both: thermal image and photography were done simultaneously by the thermographic camera from a distance of about 2 m. For further analysis, a close-up was done.

Figure 3 presents thermal images of the tattoo (part of it). The temperature is set in the range of 33.3 °C up to 36.7 °C. In the 1. case average thermal-image temperature was 35.4 °C what is higher than expected for the uncovered human skin after period of adaptation. In the 2. case, the average temperature was lower at 0.5 °C and has a value of 34.9 °C. Maximum found difference in temperatures between skin without and with tattoo were 0.4 °C in the 1. case and 0.6 °C in the 2. case. The average temperature of the inked skin of the best visible upper part of "R" letter, was for the 1. case 35.5 °C and non-inked skin that adjacent to the tattoo was 36.0 °C. For the 2. case the aforementioned values were found at the level of 35.4 °C and 35.7 °C respectively. More details are presented in the Fig. 3. Figure 4 (upper and lower) presents histograms for thermal images presented in Figs. 3 left and right respectively.

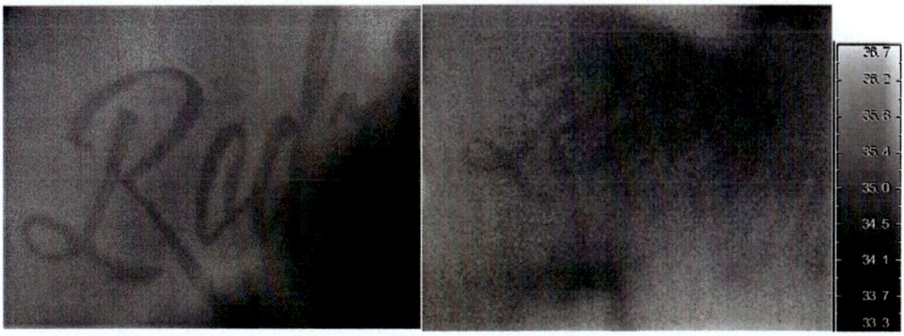

Fig. 3. Exemplary thermal images made during descripted experiment from a distance of 0.1 m with visible "Rodzi" (For print purposes a contrast and light level of the picture were adjusted, sharpening was also applied).

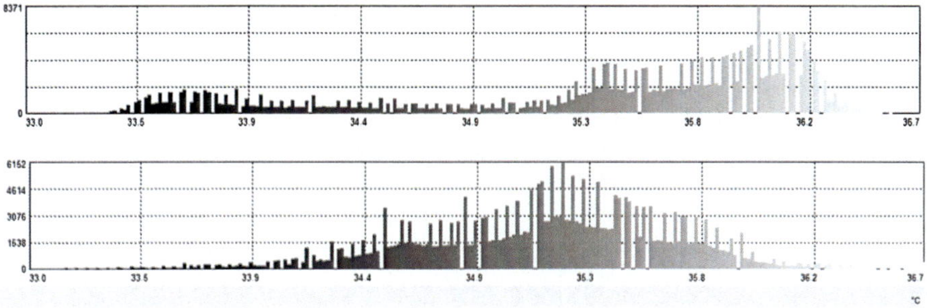

Fig. 4. Histograms of thermal images presented in Figs. 3 left (upper) and 3 right (lower).

4 Concluding Remarks

From the presented examples it can be concluded, that tattoo have to be taken under consideration during thermal body examination. It can modify temperature readings. It has to be highlighted that it is not known if the difference in registered temperatures is caused by the differences at the skin level (skin with and without ink) or is caused by different level of emissivity for inked, darker skin. This problem needs a deeper study. Nevertheless, it seems necessary for each precise examination, to check if tattoo is not visible from a short (5–20 cm) distance. From a greater distances tattoos sometimes are not detected in IR spectrum, what can be assumed, is connected with resolution of the thermal imaging cameras. As it was shown in a previous work, it is necessary to elaborate experiment with higher number of volunteers with different types of tattoo. This should allow to answer following questions: (i.) when and why the body-tattoo influences the temperature distribution and (ii.) under which conditions it is possible to detect this effect on the tattooed body. At this moment it can be supported the statement, that tattoo can influence the distribution of the skin temperature and this phenomenon can occur for relatively new and old tattoos.

Acknowledgements. Author would like to thank Mr. Łukasz Kaczorowski, who participated in the preparation of the first article from the series.

References

1. Van der Velden, E.M., et al.: Tattooing and its medical aspects. Int. J. Dermatol. **32**, 381–384 (1993)
2. Papameletiou, D., Zeni, A., Schwela, D.: Status report on the current situation, nature and size of the problem regarding the safety of tattoos, body piercing and of related practices in the EU. In: Proceedings of the Workshop on Technical/scientific and regulatory issues on the safety of tattoos, body piercing and of related practices organised by the JRC/PCE/IHCP on behalf of Directorate General for Health and Consumer Protection, pp. 9–13 (2003)
3. Based on data obtained with courtesy of the Research Group IQS (IQS Sp o.o. Francuska str. 37, 03–905 Warsaw). The study was carried out on 800 volunteers aged 15–50, https://grupaiqs.pl/ dostęp 8 VIII 20189
4. Bittencourt, M.J.S., de Parijos, A.M., da Fonseca, D.M.: Dermatofibroma in a black tattoo: report of a case. An. Bras. Dermatol. **88**(4), 614–616 (2013)
5. Zegarska, B., et al.: Complications following black henna tattoo. A case report (in Polish). Adv. Dermatol. Allergol. **6**, 278–281 (2006)
6. Bauer, J., Dereń, E.: Standardization of thermographic studies in medicine and physical therapy (in Polish). Acta Bio-Opt. Informat. Med. Inżyn. Biomed. **20**(1), 11–20 (2014)
7. Adamczyk, J.G., Boguszewski, D., Siewierski, M.: Thermographic evaluation of lactate level in capillary blood during post-exercise recovery. Kinesiology **46**(2), 186–193 (2014)
8. Moreira, D.G., et al.: Thermographic imaging in sports and exercise medicine: a Delphi study and consensus statement on the measurement of human skin temperature. J. Therm. Biol. **69**, 155–162 (2017)
9. Broussard, K.A., Harton, H.C.: Tattoo or taboo? Tattoo stigma and negative attitudes toward tattooed individuals. J. Soc. Psychol. **158**(5), 521–540 (2018)
10. Serup, J., et al.: Identification of pigments related to allergic tattoo reactions in 104 human skin biopsies. Contact Dermatitis **82**, 73–82 (2020)
11. Zagrodny, B., Kaczorowski, Ł, Awrejcewicz, J.: Influence of Body Tattoo on Thermal Image—A Case Report. In: Gzik, M., Paszenda, Z., Pietka, E., Tkacz, E., Milewski, K. (eds.) AAB 2020. AISC, vol. 1223, pp. 209–214. Springer, Cham (2021). https://doi.org/10.1007/978-3-030-52180-6_23
12. Zagrodny, B., Wojnicz, W., Ludwicki, M., Awrejcewicz, J.: Could thermal imaging supplement surface electromyography measurements for skeletal muscles? IEEE Trans. Instrum. Meas. **70**, 1–10 (2021)

Author Index

A. Hadamus et al. (Eds.): BIOMECHANICS 2021, LNNS 328, p. 159, 2022.
https://doi.org/10.1007/978-3-030-86297-8